ON DOCUMENTATION OF
SCIENTIFIC LITERATURE

ON DOCUMENTATION OF SCIENTIFIC LITERATURE .

Th. P. Loosjes
Director of Library, Agricultural University,
Wageningen, Holland
Professor of Documentary Information,
University of Amsterdam

LONDON   BUTTERWORTHS

THE BUTTERWORTH GROUP

ENGLAND
Butterworth & Co (Publishers) Ltd
London: 88 Kingsway, WC2B 6AB

AUSTRALIA
Butterworths Pty Ltd
Sydney: 586 Pacific Highway, NSW 2067
Melbourne: 343 Little Collins Street, 3000
Brisbane: 240 Queen Street, 4000

CANADA
Butterworth & Co (Canada) Ltd
Toronto: 14 Curity Avenue, 374

NEW ZEALAND
Butterworths of New Zealand Ltd
Wellington: 26-28 Waring Taylor Street, 1

SOUTH AFRICA
Butterworth & Co (South Africa) Pty Ltd
Durban: 152-154 Gale Street

Suggested U.D.C. Number 002

This edition is based on a translation by A. J. Dickson of
Dokumentation wissenschaftlicher Literatur

© BLV Verlagsgesellshaft, München - Basel - Wein

This edition
© Butterworth & Co (Publishers) Ltd, 1973

ISBN 0 408 70429 2

Printed in England by Northumberland Press Ltd, Gateshead
Bound by Richard Clay & Co Ltd, Bungay, Suffolk

AUTHOR'S NOTE

The whole book has been extensively revised and much new literature has been included. Inevitably, in a subject changing as rapidly as this, a number of important studies have appeared since the ms went to the printers. Chapters 10 and 12 have been almost entirely rewritten. The appearance of the excellent annual volumes of the *Annual Review of Information Science and Technology* may help those readers who want more elaborate information in the field of documentary information in their further studies.

CONTENTS

ONE

WHAT IS DOCUMENTATION?

THE WORD 'DOCUMENTATION'

The word 'documentation' is cognate with 'document' and 'to document'. In colloquial speech the word 'documentation' is generally used with a very wide connotation. In our subject, on the other hand, the word 'documentation' should represent a hard and fast *technical term*, and from this many difficulties arise. When a word is taken from colloquial speech into technical terminology, it must be exactly established what processes and/or things this word—here 'documentation'—should denote; thereby, the establishment of the term should fit as closely as possible to the usual, accepted, meanings of the words.

DEFINITIONS OF THE WORD 'DOCUMENTATION' IN THE LITERATURE

In the literature of documentation there is a multitude of theoretical definitions of the term. Examples may be found in the publications of van der Laan (1947), Pietsch (1954) and Verhoef (1960). In order to give a general picture, the definitions are here summarised into two main groups for discussion:

1. Definitions limited to librarianship.
2. Definitions without such limitation.

DEFINITIONS OF DOCUMENTATION LIMITED TO LIBRARIANSHIP

(a) Over-all definitions (or supraposition).

1

(b) Parallel definitions (or juxtaposition).
(c) Subordinate definitions (or infraposition).

Definitions of 'Supraposition'

'Supraposition' definitions describe the subject of documentation so exhaustively that they also include the totality of librarianship. Everything is regarded as a function of documentation so far as the acquisition, arrangement and exploitation of documents are concerned, including the work of the publisher.

In effect, by *documents* we mean all *units of material containing information*, not only written and printed matter but also films, discs and tapes. The best-known example of this type of definition is the lattice concept of Otlet (1934), which is also employed by FID and to which Frank (1949) and Briët (1951) also subscribe. It states: 'Documentation is the collection, arrangement and distribution of documents of every sort in all fields of human activity.'

Here also belongs the definition of Shera (1951), who speaks of documentation as 'bibliographic organisation' and understands by this: 'The canalisation of graphic records to all users, for all purposes and at all levels (of use) in such a way as to maximise the social utilisation of recorded human experience.'

Definitions of 'Juxtaposition'

The proposers of 'juxtaposition' definitions regard 'librarianship' and 'documentation' as parallel terms. Pietsch (1954) comes to the conclusion, by means of an international questionnaire, that librarians 'look after' and documentalists 'exploit' the collection. Fill (1954) also takes this line when he ascribes to the library the *administration* and to documentation the *exploitation* of the documents. For him the humanities are the specific field of the library, whereas documentation deals essentially with applied sciences and technology. The Documentation Committee of the Netherlands Library Association (Verslag (Report), 1944) is also of this opinion. Sometimes specific types of documents are also used as criteria for juxtaposed limitation; for example, compilation of a classified catalogue for books would be evaluated as library work, whereas the compilation of the same catalogue for articles from periodicals should come under documentation (Reeser, 1954). But where is the boundary if pamphlets are counted as books, particularly since lengthy offprints of articles from major periodicals often appear as pamphlets? Librarians differentiate between immature, transient and merely topical material and mature enduring units, and their main interest is in a certain minimum quality for the cataloguing and analysis of the documents.

Definitions of 'Infraposition'

The group of 'infraposition' definitions includes the system of Kunze (1954), who sees in documentation merely a functional extension of librarianship, the difference being solely in the intensity of extending the literature with which the librarian and documentalist are concerned. According to Björkbom (Pietsch, 1954), documentation is nothing new; in his opinion it belongs entirely to the library, particularly as such tasks are undertaken in most research libraries, if perhaps with some variation in depth. Therefore Björkbom prefers not to use the word 'documentation' at all and suggests other word formations: literature service, library service, literature review, abstract service, dissemination of information. If the concept of documentation were not already accepted, we could second this proposal, but now it is too late. Further, Björkbom (1959) has also changed his mind and stated (in paraphrase): 'Documentation is bibliography and library information work suited to the situation in special libraries.'

DEFINITIONS OF DOCUMENTATION WITHOUT LIMITATION TO LIBRARIANSHIP

Pietsch (1954) quotes two authors, Picard and Scotecci, who understand documentation to mean the collecting or a collection of documents in a particular field. The word is used in just the same way in colloquial speech. An attempt was also made to delimit documentation by using the word 'bibliography'; this definition is, however, very unsound, since 'bibliography', too, can have many meanings. An attempt was made by van Riemsdijk (1941) to make the terms 'bibliography' and 'documentation' mutually exclusive.

CONCLUSION

In view of the multitudinous definitions and the subsequent discussions, it seems better to defer any final decision and simply arrive at a working hypothesis, on the basis of the historical origins of the use of the word by subject specialists, and to see which processes can be called 'documentation'.

THE ORIGINS OF DOCUMENTATION

Prior to the invention of printing in medieval Europe, monks copied the manuscripts (MS) and the copies were passed into the abbey libraries (LI), where the scholars (RE), in those days themselves monks, evaluated these works (*Figure 1*).

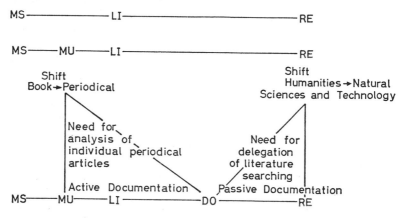

Figure 1. The rise of documentation

Only after the invention of printing was there any possibility of multiplication (MU); the products of this multiplication were passed into the libraries.

Two developments complicated this rather simple situation: first, during the seventeenth and eighteenth centuries periodicals appeared in addition to books, and, among these, the specialist journals; secondly, in the nineteenth century, in addition to the humanities, natural sciences and technology became important.

As a consequence of the displacement of the emphasis in publication from the book to the periodical, the necessity arose for a continuous analysis of articles from periodicals, particularly as scientists and technologists had to be informed speedily of the latest stage of research in their fields. This analysis soon became known as 'documentation'. The rise of the natural sciences led to literature searching becoming another duty of the documentalist.

Documentation can be said to have two starting points:

1. From the library, i.e. from the need for an analysis of individual periodical articles.

2. From research, i.e. from the necessity for delegation of literature searches.

The first is historically dealt with (displacement, book–periodical) in Chapter 2; the second (displacement, classic humanities–sciences), in Chapter 8.

The development of documentation is considered as an integral part of the business of research; however, the need for documentation arose partly outside the field of research, namely in business. Details are given in Chapter 8; for the historical dates of what could be called the prehistory and actual history of documentation, see Pietsch (1963).

POLARITY OF DOCUMENTATION

'Active documentation' originated with the requirement of individual article analysis. The so-called 'passive documentation' arose out of a need for delegation of literature searches. In active documentation the work is undertaken without the commission or specific query of an interested party. Passive documentation must be commissioned by a third party. Documentation has its two historical starting points, library and research, to thank for this polarity (see *Figure 1*). In addition, it is necessary to point to the fact that the organisational crystallisation of documentation was very much related:

(a) in the U.S.A. to reprography, as the early history of the American Society of Documentation is tied up with Watson Davis and his sponsoring of the use of microfilm (Schultz and Garwig, 1969);

(b) in Europe to the maintenance and elaboration of the U.D.C., as reflected in the early beginnings of the predecessors of the Fédération Internationale de Documentation (Rayward, 1967).

ORIGINAL WORKING HYPOTHESIS

Using the above considerations, we come logically to the working hypothesis that the field of documentation covers two different processes:

1. The analysis of articles from periodicals which can be extended when necessary to bibliographic control of *all* types of publication (active documentation).

2. The different phases of literature searching that research disciplines are ready to delegate (passive documentation).

THE FOUR-PHASE CYCLE AND ITS ECONOMIC PARALLELS

When we consider *Figure 1*, we must come to the conclusion that the grouping is not complete, since the research worker himself prepares another manuscript. Thus we may not choose a straight line as illustration but a circle (*Figure 2*). This circle covers two fields:

1. The four phases of processing: multiplication (MU), library (LI), documentation (DO) and research and investigation (RE).
2. The four forms of the material: manuscript (MS), which after multiplication becomes a publication (PU) and then library material processed (LMP), and, finally, documented material processed (DMP).

Production occurs in sectors MU and RE, distribution in sectors LI and DO and consumption in sector RE. Egan and Shera (1953) also use this representation, which is borrowed from economics.

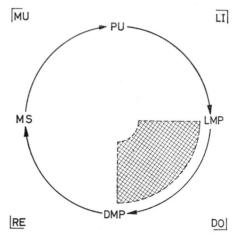

Figure 2. The interrelation between multiplication, library, documentation and research*

* This circle and the circles derived from it in subsequent pages were first published by the author in 1954 (Loosjes, 1954). The 'chain of information' of the Weinberg Report (Weinberg, 1963) is similar in concept though clad in a more modern dress.

Translator's note: Wissenschaft in the original is rendered here as research. It also appears according to context as: science, knowledge, learning, scholarship, discipline.

DOCUMENTATION AND LIBRARY

In the search for a definition which clearly separates library from documentation we have studied the origins of documentation. Its polar character is thereby clearly revealed, in accordance with our original working hypothesis. Documentation arose on the threshold of two spheres of influence, deep-rooted, on the one hand, in research work (passive documentation) and, on the other, in librarianship (active documentation).

From the history of its origins the question arises: Is documentation an autonomous discipline in relation to librarianship? In general, the boundary between the two fields can be defined as follows: As long as the *work* on an available collection of material refers to one collection, it is, regardless of any further processing, library work; if, however, it refers to only one specific subject (that is, not to the totality of one collection), it is documentation. Documentation is therefore the neighbour of research work, which indeed cannot be concerned about which stock the material finally belongs to. A good example of the dividing line between the two processes is found in the origin of many (not all) abstract journals.

A library begins to add to the normal accession lists of books and periodicals an accession list of articles in periodicals. Then annotations are added to the titles of these articles. Later on, collaboration with another library working on the same subject is undertaken and an annotated accession list of articles in periodicals is published. The decisive step follows: One no longer merely accepts the accession lists of the two founder libraries but examines other relevant and external factors and asks, what is new in this field? With this decision the step from library to documentation work is made.

In this example the boundary is easily drawn. What is the position, however, if the librarian of a special library 'documents' for himself, i.e. for the selection of material for his library? What in fact is he then doing? His action refers (negatively) to his collection; it is, however, in essence a compilation of titles in a special subject.

References

Björkbom, C. (1959). 'The History of the Word Documentation within the FID.' *Revue Docum.* **26**, No. 3, 68

Briët, S. (1951). *Qu'est-ce que la documentation?* 7th edn. Paris; U.S.O.D. Librairie des Ciseaux

Briët, S. (1954). 'Bibliothécaires et documentalistes.' *Revue Docum.* 21, No. 2, 21

Fill, K. (1954). 'Dokumentation und Bibliothekswesen.' *Nachr. Dokum.* 5, No. 3, 119

Frank, O. (1949). *Einführung in die Dokumentation.* Heft 5. Stuttgart; Dorothean Vlg

Kunze, H. (1954). 'Dokumentation und wissenschaftliche Bibliotheken.' *Dokumentation* 1, 78

Laan, A. van der (1947). 'Taak en Functie van de Documentalist.' *NIDER Publ.* No. 273

Loosjes, T. P. (1954). 'Literatuuronderzoek en Documentatiedienst.' In: *Symp. Over Literatuuronderzoek,* pp. 7–17. The Hague; *NIDER Publ.* No. 9, 2nd series

Otlet, P. (1934). *Traité de documentation.* Brussels; I.I.B.

Pietsch, E. (1954). *Grundfragen der Dokumentation,* pp. 18–39. Schriftreihe Arbgemein. Rationalisierung Landes. Nordrhein-Westfalen. Heft 14. Dortmund; Verkuhrs- und Wirtschafts-verlag

Pietsch, E. (1963). 'Die künftige Entwicklung der Dokumentation.' *Libri* 12, No. 4, 287

Rayward, W. B. (1967). 'The UDC and FID. A Historical Perspective.' *Libr. Q.* 37, No. 3, 259

Reeser, H. (1954). 'Het Inlichtingenbureau van de Universiteitsbibliotheek.' *Bibliotheekleven* 39, No. 5, 149

Riemsdijk, S. G. A. van (1941). 'Bibliografie en Documentatie. Een poging tot begripsbepaling.' *Bibliotheekleven* 26, 41

Schultz, C. K. and Garwig, P. L. (1969). 'History of the American Documentation Institute.' *Am. Docum.* 20 (April), 152–160

Shera, J. H. (1951). 'Documentation, its Scope and Limitations.' *Libr. Q.* 21, 13

Shera, J. H. and Egan, M. E. (1953). 'A Review of the Present State of Librarianship and Documentation.' In: Bradford, S. C. *Documentation.* 2nd edn. London; Crosby, Lockwood

Verhoef, M. (1960). 'Librarianship and Documentation.' *UNESCO Bull. Lib.* 14, 193

Verslag (1944). 'Verslag van de Documentatiecommissie van de Nederlandse Vereniging van Bibliothecarissen.' *Bibliotheekleven* 29, No. 9, 21

Weinberg, A. M. (1963). 'Science, Government and Information.' *A Report to the President's Science Advisory Committee.* Washington

TWO

THE DEVELOPMENT OF PERIODICALS AND THE HISTORY OF ACTIVE DOCUMENTATION

THE LITERATURE ON RESEARCH PERIODICALS

The literature on the periodical as a form of publication and as a subject in librarianship and documentation is particularly extensive. Those wishing to explore it are referred to the works of Gable (1937), Grenfell (1953), Lehmann (1936) and Osborn (1955). Reiman (1953) compiled a bibliography of the literature on periodicals, to which Tydeman (1953/4) added a supplement. A good example of a national review of the area covered by the bibliography of periodicals is the paper by Collison (1952). An index of bibliographies of periodicals arranged geographically is provided by Björkbom (1953a). In all bibliographies and literature guides special sections are devoted to periodicals and serials. The same applies to current bibliographies, abstracts journals and indexes, where new periodicals and serials are regularly listed. Periodic indexes to the literature on periodicals are to be found in, for example, *Library and Information Science Abstracts*, as well as in different abstracts sections of library periodicals and in *Bibliographic Index*. In addition, the American periodical *Serial Slants*, published from 1950 to 1956, was uniquely concerned with periodicals and, under the heading 'Serial Sources', also contained a bibliographic section on indexes to periodicals.

The Boundary between Periodical and Book

The discussion concerning the boundary between the various groups of publications is not of recent origin (see Kirchner, 1928; Malclès, 1950; or Grenfell, 1953).

In general, the book differs from the group of periodicals (series, serials, etc.) in the completion of the publication. So long as a book, even though it may be in several volumes and with many supplements, appears within a certain prescribed period within a fixed scheme, it is to be reckoned as a book, notwithstanding that in the catalogue over a period of years a so-called 'open title' is essential.

The criteria for a periodical and a series are a non-prescribed period of time and the fact that the separate parts appear under a common title. Other outward signs—for example, an editor, a certain periodicity—are not so decisive. There exist publishers' series without official mention of editorship, although most of these, behind the scenes, naturally call upon a few consultants at least as editors. There are also series in which the periodicity varies or is not at all evident.

It should also be remembered that a publication with a predetermined arrangement complete in itself represents a unified presentation of the subject, but that such a presentation can never be the object of a series or a periodical.

Although this boundary proves very useful in practice, there are transitional forms. For example, the title *Nova Guinea* began as a single report on a particular expedition to New Guinea; its publication finally stretched over many years. In time there appeared in this same series reports on other expeditions to the same area and logically the first issue was continued as a periodical. The reverse (from periodical to book) also occurs: *Die mykologischen Hefte*, edited by Kunze and Schmidt (Leipzig), commenced as a periodical (No. 1, 1817); on the appearance of No. 2 in 1823, the authors announced that they had abandoned the periodical style and that the second issue was to be counted as the second volume of a book. Kirchner (1928) was able to establish that works appearing by instalments in earlier times were often entered in lists of periodicals.

HISTORY AND DEVELOPMENT OF THE PERIODICAL

In view of the volume of literature on the history of periodicals, only a few starting points for further study can be given here. The numerous publications on the history of various individual periodicals are not considered. Many references on the rise of the learned journals are given by Ornstein (1928), Kronick (1962), Porter (1964), McKie (1948) and Kirchner (1958–62). Garrison (1934) remarks that the mother of the learned journal can be taken to be the learned society; and its father, the newspaper. Indeed, the two main sources of material of the first journals were, on the one hand, the papers read by members to meetings of the learned societies and, on the other, the 'News'. According to Harff (1941), irregular serial publications by members of learned societies were often precursors of learned journals. It is generally accepted that the *Journal des Sçavans* published on 5th January 1665 was the first learned journal, although periodical-like precursors can be traced with certainty (*German Book Catalogues* and the *Gesta Lyceorum Accademia dei Lincei 1609*). The founder of the *Journal des Sçavans*, Denis de Sallo, started his work by having references copied out for himself and, later, for other interested persons. In addition, he gathered book reviews, news of scholars (personalia), reviews of the work of researchers (dilettanti), publications of new results of experiments and judicial judgments.

Three months after the *Journal des Sçavans* appeared, the first of the *Philosophical Transactions of the Royal Society* was published in London. According to McKie (1948), the *Journal* provided the springboard for the London publication. The *Transactions* provide the classic example of the learned journal, as we know it today; the *Journal des Sçavans* aimed at a larger readership. Harff (1941) notes that the learned journals from the very beginning eschewed popularisation.

The two above-mentioned periodicals were joined three years later, in 1668, by the *Giornali di Letterati*, which is described by Ornstein (1928) as an imitation of the *Journal des Sçavans*. The third original learned journal was the German *Miscellanea Curiosa*, a publication of the Collegium Naturae Curiosum. It used the *Philosophical Transactions* as a model and is the oldest periodical which specialised in particular subjects.

That periodicals were flourishing at an early date can be inferred from a remark by Henkle (1951), according to whom there was a protest in 1841 against the great mass (43!) of *German* medical periodicals. According to Brodman (1954), there were, globally, 910 medical periodicals in the year 1900; *World Medical Periodicals* (3rd edn) reported in 1961 over 5800 titles. That the periodical in the course of its development gradually lost its news character can be seen from the frequency statistics of Lorenz (1937) in Table 1. (The figures represent rounded percentages of the total numbers of periodicals.)

Moving away from the dissemination of news, the periodical gradually took over the publication function of the book, to which

Table 1 (After Lorenz, 1937)

Frequency	1700–1800	1850	1902	1910	1926	1932
Weekly	85	29	28.4	34	25.5	22.7
Bi-weekly	0.5		17.8		18.1	17.1
Monthly	3.8	25	23.2	40	38.4	36.4

increasingly fell the task of recapitulation. A good example of this is given by Koumans (1954) in his literature search on the toxicity of penicillin: up to and including 1947 there had appeared 108 articles in periodicals on this subject, whereas it was not until 1948 that the subject was first dealt with in book form. In a review of *Chemical Abstracts* and its organisation it was observed that books do not usually contain results of new research (Anon., 1951). The news function has been taken over by features such as *Letters to the Editor*, as well as by typical priority periodicals—for example, *Nature*.

QUANTITATIVE RATIO OF BOOK TO PERIODICAL LITERATURE

According to Osborn (1954), 75 per cent of the material being received by the Library of Congress is of 'serial character'. Rogers and Adams (1950) give figures on the shift of emphasis from book to periodical: they counted the monographs and articles from periodicals, respectively, which were received from the first series (1872) up to the fourth series, Vol. 10 (1948) inclusive, according to the contents by subject in the *Index Catalogue*. This study showed that, initially, 75 per cent consisted of periodical material, and that in 1948 it was 92 per cent. According to Grosser (1948), requisition slips received by the Swiss Union Catalogue for periodicals previously represented 5 per cent of the demand; in 1948, 30 per cent. Egger (1956) reports from the same catalogue that 50 per cent of the queries for post-war literature referred to articles from periodicals. However, this was valid mainly for the natural sciences. Stevens (1956), comparing the humanities and natural sciences in this respect, states that 90 per cent of the references in the literature of physics and chemistry and 85 per cent of the references in geology are to articles from periodicals, quoting Fussler (1949) and Gross and Woodford (1931), respectively, as his authorities. Against this, only 9 per cent of the literature of the history of the U.S.A. (McAnally, 1951) and 46 per cent of the literature of sociology (Broadus, 1952) refers to articles from periodicals. Egger (1956) mentions that the humanities account for 80 per cent of the *book* requisitions to the Swiss Union Catalogue but only 30 per cent of periodical requisitions, whereas natural sciences, medicine and technology account for 70 per cent. Brown (1956) found in a reference count in the literature of chemistry and in the literature of mathematics 93 per cent and 76 per cent, respectively, of references were to periodicals.

CAUSES OF THE SHIFT OF EMPHASIS FROM BOOK TO PERIODICAL

The shift of emphasis from book to periodical, in the publication of the results of research, can be attributed to the following causes.

Social Causes

A publication is important in establishing priority. Therefore publications tend to become shorter and follow one another more rapidly as the competition becomes more intense. In addition to their normal function of dissemination of information, publications have nowadays to fulfil a further assignment: they must serve the author as a 'means of advancement'. Herner (1954) and Richardson (1951) both emphasise that frequent publications (1) raise income levels, (2) make a name in professional circles and (3) improve career prospects for the author. The steadily rising flood of publications is therefore no longer to be traced solely to the increased need to inform others. In fact, Richardson even states the 'desirable' maximum and minimum number of publications:

> Have you never met the chemist, who has set himself the target of ten per annum or one monthly publication? Such a person is an ambitious realist, who takes life as it is. Without this target it may well have taken five or ten years longer to have become a professor. Have you the heart to damn him? I could almost (but not quite) forgive him if he were to make two publications out of one in republishing data already communicated, which would be all the same an all too-common practice, or that he would split an article complete in itself into two unnecessarily ...

The American Chemical Society Professional Education Committee publicly suggested a minimum of 10 publications within 19 years for a top-rank university teacher and that a thorough, critical examination of all the publications be required of a teacher who published less. That quality may well be greatly impaired in this way can be seen from the information given by Bourgeois (1956) that out of 100 technical articles only 8 made a really original contribution to learning and research. Meyerhof (1961) tells of a letter received by the editor of *Science*, in which an author requested publication in December instead of January as planned, because ... it would not count towards his total of publications for the year; this figure determined the increments and promotions in the institute concerned. (See also Chapter 3, p. 23.)

Economic Causes

The investment costs chargeable to a contract, costs of distribution, and publicity and general overheads have risen steeply, thus making single editions even more expensive and risky and driving publishers even further towards series.

Technical Causes

The economics of using improved and faster-working printing machines demands ever-higher printing numbers. Coupled with this fact is the commercial risk involved in the production of a large-scale publication. This results in small units, such as articles, being grouped together in periodicals. A subscription for a periodical is a good example of an 'imposed condition of sale'. This situation is somewhat improved by the so-called repackaging journals: the original material published in a scientific journal is repackaged and distributed for special groups of subscribers. This method has been made possible largely by the use of computer-controlled type-setting.

COUNTERCURRENTS

There are two traceable countercurrents to this shift of emphasis from book to periodical, one artificial and one (ever-increasing) natural. The first countercurrent finds expression in the efforts of research workers to rid themselves of an imposed condition of sale. New suggestions are being repeatedly made in this direction, since the demand for reprints of particular articles and even for single issues of periodicals, not within the subscription, is constantly growing. A summary of all the suggestions for reform in modern research periodical production is given by Phelps and Herling (1960). The publications of Björkbom (1953b), Woke (1951) and Laclémandière (1951) are also well worth studying. The long list of authors quoted by Phelps and Herling is an indication of the serious intent of these efforts. Publishers are reproached throughout for an imposition of condition of sale. Subscribers have to accept willy-nilly a whole series of other articles just for the sake of individual papers. (In the case of extremely specialised periodicals this generally applies to a lesser extent.)

'Publish each article separately!' This advice is simple but it overlooks the fact that the combination of a number of articles makes the marketing of single articles, at a reasonable price, at all possible; the price of reprints is purely a 'run on' price, since its set-up costs are contained in the setting costs for the whole periodical and are borne by all the subscribers together. Moreover the efficient distribution of these separates proved to be a problem: authors didn't order them (Bernal, 1965).

With batch production of each single article the setting costs would raise the retail price well over the normal prices for offprints. Reprints live as it were parasitically on the host journal. In addition to this, only a periodical can really undertake to publish occasionally a valuable article with a restricted readership, just because it 'gets by' in the crowd. There could be no question of this in the system of batch production as suggested; articles of interest to only a few readers would simply not be published. Be that as it may, the view that imposition of a condition of sale could be avoided dies hard. Pownall (1926) has devoted a whole book to its advocacy. Again and again the old suggestions are trotted out, as the report of Coblans (1957) shows. Kyle (1957) sees the possibility of realising the wish of research workers, that budgetary authorities, e.g. foundations or the public purse, should make the granting of funds to publishers dependent on their making large quantities of reprints available to international exchange centres.

The second countercurrent stems from the technical development of reproduction processes covered by the expression 'office printing machines' (stencil, office offset machines, etc.). Obviously, this development will continue; and already technology offers immense possibilities for short runs and quick production. It is here that the economic argument, and the argument on technical grounds, for the shift of emphasis from book to periodical fail. Against this, the social causes remain a potent factor, for so long as the appearance of the more simple and cheap reproductions, with typewriter fount, no right-hand margin justification, etc., lags significantly behind that of letterpress, it will remain a question of status to have one's manuscripts somewhere, somehow, printed in fitting form. Moreover, the form of the printed periodical suggests a more searching refereeing of material, whereas many readers find that the office copy with its more simple exterior does not produce the same effect of depth and editorship. In the same context, however, as the so-called 'near-print' process further develops, so will reader resistance on the aesthetic plane certainly diminish and so, too, will the discrimination become blurred or disappear entirely. As yet, however, even this second countercurrent has remained without influence on the major line of development, still extending in the direction of ever greater shift of emphasis from book to periodical as a medium for scholarly publication. This does not mean that

the scientific periodical, as we have it now, is not regularly scrutinised for efficiency as a vehicle for distribution or research results.
A recent example of this type of scrutiny is the study of Herschman
(1970), who discusses the value of the scientific journal compared
with the newer techniques for information dissemination.

The main stream of criticism of the scientific journal stems from
the ever-growing time lag between the acceptance of the manuscript
and its publication in the journal. The author's natural impatience
has led to a number of additional publication channels supplementing, each in its own way, the classic channel of the scientific
periodical. Among these additional channels may be mentioned:

(a) Technical reports. Long before final publication in the journals
interim and progress reports with a limited circulation (and,
hence, very difficult to trace and to obtain) accompany the
progress of a research project. This literature is a problem
area for every librarian and documentalist. Separate bibliographic tools have even been created, either in the form of
current bibliographies of these reports or in the form of
periodical reviews about current research in the different
kinds of agencies.

(b) Manuscript depositories with announcement-bulletins with
abstracts of the deposited material. This system is reluctantly
accepted by authors, for understandable reasons. Sorokin
(1969) describes the manuscript depot of a Russian journal in
the field of atomic energy. More acceptable to authors is a
partial publishing–partial depositing (especially of cumula-
tions of data) method practised in Canadian scientific
periodicals.

(c) Special additional bulletins for immediate publication of existing scientific journals, e.g. *Tetrahedron Letters*, created by
the periodical *Tetrahedron*. Kuney (1970) reports that 50 per
cent of the letters sent to *Physics Reviews Letters* never arrive
subsequently as a normal article for *Physics Reviews* itself.

(d) Organised preprint circulation. This short-lived experiment
was inaugurated by the U.S. National Institutes of Health.
Under the sponsorship of this organisation there were during
1961–1966 seven so-called Information Exchange Groups in
different disciplines. The members of these groups had an
internal exchange organisation of preprints. This experiment
was ended when six leading publishers of scientific journals
announced a boycott of all members of these groups (Webb,
1970). The success of the experiment was considerable;
Coblans (1970) reports an output of $1\frac{1}{2}$ million preprints.

Understandably, scientists wished to see further development in this direction (Bever, 1969).

Garvey and Griffiths (1967) have thoroughly studied the publication delays in scientific periodicals in the field of psychology. They recommend as eventual means for acceleration of the flow of information:

1. Preconvention published proceedings.
2. Accelerated flow of the material (even before publishing) to the abstracts journal of the field in question.
3. Listing in the journal the accepted manuscripts awaiting publication.

THE LIBRARY AND THE GROWING NEED FOR AN ANALYSIS OF PERIODICAL ARTICLES

In Chapter 1 it was noted that the shift of emphasis from book to periodical gave rise to the need for an analysis of individual periodical articles: of special interest in this connection is the review paper by Shera and Egan (1953). They describe at length the attitude taken by library circles towards this need. With the development of the research periodical beginning in mid-nineteenth century, librarians doubtless felt that the analysis of individual papers in periodicals fell within their province and they also discussed at their various conferences the difficulties that this analysis would entail. However, when the exponential development of the periodical began, libraries were lacking in staff and funds to face this major new task. It was indeed soon recognised that the contents of the periodicals were the same for all libraries and that cooperation was desirable. Despite this, a centralised index service as such did not materialise: libraries were not sufficiently well organised to begin such an undertaking and to see it through successfully. In the case of periodicals, librarians continued in their traditional role of conservationists and therefore other centres had to undertake the task of exploiting the periodicals. It was soon apparent that this was not to the advantage of the development of librarianship, which is all the more to be regretted, because in the early stages the librarians of that time doubtless saw in this a job to be tackled. The pioneers of modern librarianship, who are named in detail by Egan and Shera, e.g. Edwards and Panizzi in England and Jewett, Winsor, Cutter and Poole in the U.S.A., saw full well that it was the business of librarians to provide access to periodicals as well. At that time, however, librarians were more concerned with the pedagogic aspect of their calling: they had emblazoned mass education on their banner and therefore they did not have the time (or indeed the enthusiasm) when the flood of periodicals rendered imperative a systematic evaluation, beyond the call of library control and storage.

Since in this respect there was not much to be expected from libraries, the documentalists took it upon themselves *circa* 1900 to analyse the contents of periodicals. From then on, according to Egan and Shera, the paths of librarianship and documentation diverge further and further, since there is a major difference between a collection serving pedagogic ends and one giving exact answers to exact questions from science and research. Therefore in librarianship itself a split occurs into public and special libraries: the public libraries remain with their ideal of education and the special libraries concentrate on their analytical function. An intermediate position is taken by university libraries, which seek to combine in themselves elements of both types. The pedagogic ideal still persists ('browsing room') but the impossibility of keeping up-to-date in fields of research forces even university libraries even further to specialise: self-imposed limitations are accepted and contact and division of work with other libraries are sought (the old Metropolitan Special Collections scheme in London and the Farmington Plan in the U.S.A.) or special institute libraries are formed. The two trends usually go hand in hand.

The dichotomy has extended even into the professional organisations. Both the American Society of Information Science and the Special Libraries Association exist separately from the American Library Association. In England ASLIB is distinct from the Library Association. In the Netherlands it is true that there is, on the one hand, the Nederlandse Vereniging van Bibliothecarissen and, on the other hand, the Nederlands Orgaan ter Bevordering van de Informatieverzorging (NOBIN)—formerly the Nederlands Instituut voor Documentatie en Registratuur (NIDER)—but there a fortunate position prevails, inasmuch as the Association has set up a special libraries section. In Germany there are the Deutsche Gesellschaft für Dokumentation and the Verein deutscher Bibliothekare.

References

Anonymous (1951). 'Priestley House: Shrine of American Chemistry.' *Chem. Engng News* **29**, 3244

Bernal, J. D. (1965). 'Summary Papers and Summary Journals.' *J. Docum.* **21**, No. 2, 122

Bever, A. T. (1969). 'The Duality of Quick and Archival Communication.' *J. Chem. Doc.* **9**, No. 1, 3

Björkbom, C. (1953a). 'Bibliographical Tools for Control of Current Periodicals.' *Revue Docum.* **20**, No. 1, 19
Björkbom, C. (1953b). 'De Vetenskapliga Tidskrifternas Probleme.' *Svenska Vägför. Tidskr.* **40**, 559
Bourgeois, P. (1956). 'L'avenir du Périodique Scientifique.' *Libri* 7, No. 1, 71
Broadus, R. N. (1952). 'An Analysis of Literature Cited in the American Sociological Review.' *Am. Sociol. Rev.* **17**, 355
Brodman, E. (1954). *The Development of Medical Bibliography.* New York; Medical Library Association
Brown, C. H. (1956). 'Scientific Serials.' *A.C.R.L. Monographs* No. 16
Coblans, H. (1957). 'New Methods and Techniques for the Communication of Knowledge.' *UNESCO Bull. Libr.* **11**, No. 7, 154
Coblans, H. (1970). Book review (S. Passman, *Scientific and Technical Communication*). *UNESCO Bull. Libr.* **24**, No. 4, 216
Collison, R. L. (1952). 'Recent Developments in British Serials.' *Serial Slants* **3**, No. 2, 35
Egger, E. (1956). 'Gesamtkataloge. Aufbau und Organisation eines Gesamtkatalogs im Hinblick auf die Benützung.' *Libri* **6**, No. 2, 97
Fussler, H. H. (1949). 'Characteristics of the Research Literature Used by Chemists and Physicists in the U.S.A.' *Libr. Q.* **19**, No. 19, 119
Gable, J. H. (1937). *Manual of Serials Work.* Chicago; A.L.A.
Garrison, F. H. (1934). 'The Medical and Scientific Periodicals of the 17th and 18th Centuries.' *Bull. Inst. Hist. Med. Johns Hopkins Univ.* **2**, No. 5, 285
Garvey, W. D. and Griffiths, B. C. (1967). 'Communication in a Science: the System and its Modification.' In: de Reuck, A. and Knight, J. (eds). *Communication in Science*, pp. 16–36. London; Churchill
Grenfell, D. (1953) *Periodicals and Serials. Their Treatment in Special Libraries.* London; ASLIB (2nd edn, 1965)
Gross, P. L. K. and Woodford, A. O. (1931). 'Serial Literature Used by American Geologists.' *Science, N.Y.* **73**, 660
Grosser, H. (1948). 'Die 4. Auflage des Verzeichnisses ausländischer Zeitschriften in schweizerischen Bibliotheken.' *Nachr. Ver. Schweiz. Bibliothek.* **24**, No. 1, 11
Harff, H. (1941). *Die Entwicklung der deutschen chemischen Fachzeitschrift. Ein Beitrag zur Wesensbestimmung der wissenschaftlichen Fachzeitschrift.* Berlin; Verlag Chemie
Henkle, H. H. (1951). 'The Natural Sciences: Characteristics of the Literature, Problems of Use, and Bibliographical Organization in the Field.' In: *Bibliographical Organization.* pp. 140–160. Chicago; Univ. of Chicago Press.
Herner. S. (1954). 'Information Gathering Habits of Workers in Pure and Applied Science.' *Ind. Engng Chem.* **46**, 229
Herschman, A. (1970). 'The Primary Journal. Past, Present and Future.' *J. Chem. Doc.* **10**, No. 1, 37
Kirchner, J. (1928). *Die Grundlagen des deutschen Zeitschriftenwesens.* Leipzig; Hiersemann
Kirchner, J. (1958–62). *Das deutsche Zeitschriftenwesen. Seine Geschichte und seine Probleme.* Wiesbaden; Harrassowitz
Koumans, E. P. (1954). 'Documentation in a Medical Library.' *Libri* **3**, 295
Kronick, D. A. (1962). *History of Scientific and Technical Periodicals. The*

Origin and Development of the Scientific and Technical Press. 1665–1790.
New York; Scarecrow Press
Kuney, J. H. (1970). 'New Developments in Primary Journal Publication.'
J. Chem. Doc. **10**, No. 1, 43
Kyle, B. (1957). 'Current Documentation Topics and Their Relevance to
Social Science Literature.' *Revue Docum.* **24**, No. 3, 107
Laclémandière, J. de (1951). 'Les publications périodiques répondent-elles
aux besoins documentaires?' *ABCD* **2**, 51
Lehmann, E. H. (1936). *Einführung in die Zeitschriftenkunde.* Leipzig;
Hiersemann
Lorenz, E. (1937). *Die Entwicklung des deutschen Zeitschriftenwesens.* Berlin
McAnally, A. M. (1951). 'Characteristics of Materials Used in Research
in U.S. History.' Unpublished Ph.D. Thesis, Graduate Library School,
Univ. of Chicago. (Cited in Stevens, 1956)
McKie, D. (1948). 'The Scientific Periodical from 1665 to 1798.' *Phil.
Mag.* (July), 122
Malclès, L. N. (1950). *Les sources du travail bibliographique.* Genève;
E. Droz
Meyerhof, H. A. (1961). 'Useless Publication.' *Bull. Atom. Scient.* **17**, No. 3,
92
Ornstein, M. (1928). *The Role of Scientific Societies in the 17th Century.*
Chicago; Univ. of Chicago Press
Osborn, A. D. (1954). 'The Future of the Union List of Serials.' *Coll. Res.
Libr.* **15**, 1, 26, 118
Osborn, A. D. (1955). *Serial Publications.* Chicago; A.L.A.
Phelps, R. H. and Herling, J. P. (1960). 'Alternatives to the Scientific
Periodical.' *UNESCO Bull. Libr.* **14**, No. 2, 61
Porter, J. R. (1964). 'The Scientific Journal—300th Anniversary.' *Bact. Rev.*
28, No. 3, 211
Pownall, J. F. (1926). *Organised Publication.* London; Elliot Stock
Reiman, F. M. (1953). 'A Selected Bibliography of Articles Related to
Serials.' *Serial Slants* **4**, No. 2, 49
Richardson, A. S. (1951). 'Foundations of Chemistry.' *Chem. Engng News*
29, 2134
Rogers, F. B. and Adams, S. (1950). 'The Army Medical Library's Publica-
tion Program.' *Tex. Rep. Biol. Med.* **8**, 271
Shera, J. H. and Egan, M. E. (1953). 'A Review of the Present State of
Librarianship and Documentation.' In: Bradford, S. C., *Documentation.*
2nd edn. London; Crosby, Lockwood
Sorokin, Y. N. (1969). 'On the Possible Prospects of Primary and Secondary
Scientific Publications.' *Nauch-Tekh. Inf.* **1** (1), 3
Stevens, R. E. (1956). 'The Study of Research Use of Libraries.' *Libr. Q.*
26, 41
Tydeman, J. F. (1953 and 1954). 'Articles of Interest to Serial Librarians.'
Serial Slants **4**, No. 4, 181; **5**, No. 2, 73
Webb, E. C. (1970). 'Communication in Biochemistry.' *Nature, Lond.* **225**,
132
Woke, P. A. (1951). 'Considerations on Utilisation of Scientific Literature.'
Science, N.Y. **116**, 13/4, 339

THREE

GENERALITIES ON ACTIVE DOCUMENTATION

INTRODUCTION

Bibliographic control of articles in periodicals is a major problem in itself; however, it seems desirable to bring the bibliographic control of periodicals and serials, as well as of individual publications, into our purview. We first consider briefly the flood of literature as such—which has of course given rise to the whole problem—and the attempts to stem the tide.

OPERATION DELUGE

A few statistics should show the necessity of 'Operation Deluge' (as a librarian of the American Navy Department named it) in the face of the exponential growth of the literature: Schultze (1950) ascertained that in the last 500 years 12 million different books (titles, not copies) have been published, more than 10 million of them in the last 100 years. In the Weinberg Report (1963) it is mentioned that *Chemical Abstracts* in 1930 contained 54 000 abstracts and in 1962 some 165 000. In 1970 an increase to over 200 000 abstracts a year was reported. The 'big four'—*Chemical Abstracts, Biological Abstracts, Excerpta Medica* and **MEDLARS** (Medical Literature Analysis and Retrieval System)—have all passed the 200 000 mark. Shilling (1963) presents a telling diagram of the growth curves of periodicals and secondary abstracts journals (*Figure 3*). Bourne (1962) has compiled extensive quantitative data on the deluge of literature. His documentation is most thorough. His total for the natural sciences and technology is approximately 41 000 periodicals with 1 000 000 articles; all other disciplines he estimates at another 1 000 000 articles. He also gives data on the extent of bibliographic control.

21

Detailed quantitative data on current output of scientific and technical serials can be found in the work by Gottschalk and Desmond (1963) and in that of Gribbin (1960), who refers to measurements of literature growth in the fields of chemistry, geology and botany.

According to an outline by Clapp (1951), during the years 1899–1949 the population of the U.S.A doubled, the number of copies deposited by publishers, according to copyright law, trebled and the number of all published items increased sevenfold.

Kent (1971), writing about what he calls 'the information explosion', says:

The problem has three dimensions of frustration:

1. The impossibility of an individual reading and remembering all of the literature that has a reasonable probability of being of use later.
2. The economic impossibility of individuals or their organisations processing for later retrieval the majority of literature of possible pertinent interest.

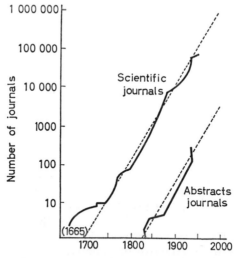

Figure 3. Growth curve of scientific journals and abstracts journals. One thing has to be borne in mind with those growth curves: there is a fair amount of discontinuation among periodicals and serials. Gottschalk and Desmond (1963) found in analysing the World List of Scientific Periodicals *a death rate of 1/3 of the titles given; a sample of* New Serials *reached* 40% (*From Shilling, 1963, by courtesy of* American Documentation)

3. The inadequacy of traditional library methods and tools in coping effectively with the detailed requirements of individuals in identifying information pertinent to a given problem.

INTERNATIONAL DISCUSSION

There has, of course, been considerable international discussion of the problem presented by the flood of literature, especially during the period since the Second World War. The history of this discussion can be found in, e.g., the paper by Ditmas (1948), who surveys the position from Otlet to the well-known conference of the Royal Society Scientific Information Conference of 1948. The proceedings of this conference contain much worth-while material on the problem, as does a follow-up Conference on Scientific Information held in Washington in 1958. Surveys of particular interest have been conducted by Clapp and Murrah (1955), in the form of a preliminary report by Coblans (1955), and Murrah (1951) gives a very concise outline well worth reading. Taube (1951) points out in detail that, in his opinion, international and national general bibliographies have no value for the research worker and that only subject bibliographies of the separate disciplines can offer a solution. The deliberations of UNESCO bodies on this complex of questions are treated in the works by Carter (1951) and Brownson (1952). In general, their object is first of all to create national order in bibliographic control before undertaking international work. This would mean, however, that work would commence on an international front only when every country had clarified the whole matter internally. In contrast, Holmstrom (1955) favours international bibliographic work without waiting for a sound national infrastructure. Special mention should be made of the support which UNESCO is giving to various activities in this field, as evidenced by the aid given to such organisations as IFLA, FID and ICSU.

One of the reasons for the deluge of literature is undoubtedly the fact that publications are made not only for communication purposes but also for priority and prestige reasons. The growing tendency towards anonymous publishing by research teams or laboratories as such may lead to a future decline of these prestige publications. Ziman (1969) speaks in this context about 'neurotic overcompetitive tendencies'. London (1968) and Bergen (1966) argue that the publication explosion is not concurrent with a knowledge explosion but rather implies that we should speak of a publication inflation (see also Chapter 2, p. 13). Ziman (1970) made a thorough analysis of the crisis situation, which he calls 'the pollution of information environment by vast quantities of waste words'. He relates the difficulties to the crisis of overpopulation, which has given rise to the debasement of quality criteria of publication; industrialisation and bureaucratisation of science; and the shift from reader-control to producer-control.

In this situation selection devices become of ever-growing importance. We think in this context of critical reviews and of the fact of the concentration of important research material in a very small number of first-rate periodicals. This qualitative concentration and its implications for literature searching are elaborated in Chapter 9.

CRITERIA OF ARRANGEMENT FOR THE DISCUSSION OF BIBLIOGRAPHIC AIDS

Before we discuss, point by point, the various aids to bibliographic control, we must be clear as to the criteria of arrangement.

The aids can be variously arranged:

1. According to static compilations complete in themselves (e.g. index of annual sets of periodicals) and according to steadily growing material (e.g. a current documentation list).
2. According to collections (a catalogue then being initiated) and according to subject content (this is documentation in the full meaning of the word). In fact, bibliographic control on behalf of a collection depends primarily on the documentation in the subject in which the library specialises; of course, this kind of documentation must precede the selection. This bibliographic control proceeds by means of the desiderata file and (as a derivative) in the catalogue.
3. According to the type of material, and this will be accepted here as the standard. Later in the book we shall discuss in due order the bibliographic control of articles in periodicals (Chapter 4), of periodicals and serials themselves (Chapter 5) and, finally, of individual publications (Chapter 6).

BIBLIOGRAPHY OF BIBLIOGRAPHIES

Yet a second observation should be considered as a premise. There is a growing group of publications, which in their way are 'bibliographies squared', i.e. bibliographies of bibliographies, whereby two types are to be distinguished: title only and title with annotation. A good example of the first type is the work by Besterman (1950); and of the second, the book by Winchell (1951). In addition, there is an imposing rank of annotated bibliographies in the form of specialised guides to the literature, available now for most of the natural sciences disciplines. However, more and more of these guides are now also appearing in the fields of history, history of art, and social sciences. There are, in addition, current bibliographies of bibliographies; for example, a section of the *Quarterly Bulletin of the International Association of Agricultural Librarians and Documentalists* is devoted to bibliographies of agriculture. Finally there are 'bibliographies cubed', of which the work by Josephson (1901), *Bibliographies of Bibliographies*, serves as an example.

References

Bergen, D. (1966). 'Implications of General Systems Theory for Librarianship and Higher Education.' *Coll. Res. Libr.* **27**, No. 5, 358

Besterman, Th. (1950). *A World Bibliography of Bibliographies*. Geneva; Societas Bibliographica. (4th edn, 1965–6. Lausanne; Societas Bibliographica)

Bourne, C. P. (1962). 'The World Technical Journal Literature: An Estimate of Volume Origin, Language, Field, Indexing and Abstracting.' *Am. Docum.* **13**, 2, 159

Brownson, Helen L. (1952). 'Recommendations and Results of International Conferences on Scientific Information and Bibliographic Services.' *Am. Docum.* **3**, No. 1, 29

Carter, E. J. (1951). 'A Survey of Achievements and Problems in National and International Bibliography.' *ASLIB Proc.* **3**, No. 4, 253

Clapp, V. W. (1951). 'The Role of Bibliographical Organization in Contemporary Civilization.' In: *Bibliographical Organization*. pp. 3–23. Chicago; Univ. of Chicago Press

Clapp, V. W. and Murrah, K. O. (1955). 'The Improvement of Bibliographic Organization.' *Libr. Q.* **25**, No. 1, 91

Coblans, H. (1955). 'Bibliography, International, National and Special.' In: *Congr. Int. Bibl. Centres Docum.* Vol. 1, *Rapports Préliminaires*. pp. 44–47. La Haye; Nijholt

Ditmas, E. M. R. (1948). 'Coordination of Information. A Survey of Schemes put Forward in the Last Fifty Years.' *J. Docum.* **3**, 209

Gottschalk, C. M. and Desmond, W. F. (1963). 'World Wide Census of Scientific and Technical Serials.' *Am. Docum.* **14**, No. 3, 188

Gribbin, J. H. (1960). 'Relationships in the Pattern of Bibliographical Devices.' *Libr. Q.* **30**, No. 2, 130

Holmstrom, J. E. (1955). 'Cooperation envisagé pour répertorier les périodiques scientifiques et signaler les bulletins de comptes-rendus analytiques où ils sont analysés.' *UNESCO NS. 490540* (280)

Josephson, A. G. S. (1901). *Bibliographies of Bibliographies*. 2nd edn.

Chicago: *Bull. Bibliogr. Soc. Am.* 1910–1912, and *Papers* of this society 1912–1913

Kent, A. (1971). *Information Analysis and Retrieval.* New York; Becker and Hayes

London, G. (1968). 'The Publication Inflation.' *Am. Docum.* **19**, No. 2, 137

Murrah, K. O. (1951). 'History of Some Attempts to Organize Bibliography Internationally.' In: *Bibliographical Organization.* pp. 24–53. Chicago; Univ. of Chicago Press

Schultze, R. S. (1950). 'Literaturflut und mechanisierte Auskunftserteilung unter Berücksichtigung der modernen Einrichtungen des Auslandes.' *Nachr. Dokum.* **1**, 10

Shilling, C. W. (1963). 'Requirements for a Scientific Mission-Oriented Information Center.' *Am. Docum.* **14**, No. 1, 49

Taube, M. (1951). 'Functional Approach to Bibliographic Organization: A Critique and a Proposal.' In: *Bibliographical Organization.* pp. 57–71. Chicago; Univ. of Chicago Press

Weinberg, A. M. (1963). 'Science, Government and Information.' *A Report to the President's Science Adv. Centre.* Washington

Winchell, C. M. (1951). *Guide to Reference Books.* Chicago; A.L.A.

Ziman, J. M. (1969). 'Information, Communication, Knowledge.' *Nature, Lond.* **224**, 318

Ziman, J. M. (1970). 'The Light of Knowledge. New Lamps for Old.' *Aslib Proc.* **22**, No. 5, 186

FOUR

BIBLIOGRAPHIC CONTROL OF ARTICLES IN PERIODICALS

BIBLIOGRAPHIC MATERIAL FOR CONTROL OF ARTICLES IN PERIODICALS

The first question that arises is: What material is available? On the basis of the circle in *Figure 2*, we can look into the question of what contributions the various sectors can make in bibliographic material (see *Figure 4*).

Multiplication Sector (MU)

1. Subject index of separate volumes of periodicals.
2. Cumulative subject index of a certain set of volumes of the periodical concerned.
3. Common index of a number of periodicals (mostly called 'index').

This material is important in itself but is only of secondary importance for bibliographic control, since in the first two cases the content of *one* periodical only is covered and in the third case the content of a *limited* number only is covered.

Library Sector (LI)

Small special libraries prepare catalogues and, sometimes, accession lists which also mention the periodical articles. Such libraries often collaborate in this activity. Some of the accession lists also contain annotations; but with this material very little progress can be made, since one is dependent upon the selection policy and budget of these libraries. There is a certain interaction between library and

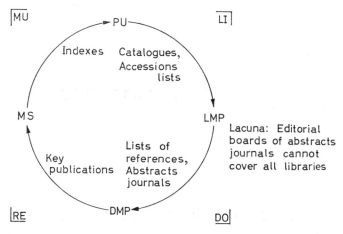

Figure 4. The contributions of bibliographic material from the various sectors to bibliographic control of articles in periodicals*

documentation activity: in the library sector documentation grows in certain circumstances from the catalogue. The reverse also takes place: in a field in which ample and good documentation is commercially available a special library can on occasion reduce the amount of catalogue work. On this interaction see, for example, Rogers (1964). Quartz (1953) even recommends a much more intensive use of abstracts journals and advises against setting up classified catalogues in one's own library.

Another kind of interaction is that whereby a library adapts its periodical subscription list to the list of periodicals covered by an abstracts journal in its field. Also, libraries may adapt their standards to the rules of the abstracts journals (e.g. transcription, citation of references, abbreviations of periodical titles). Quite another aspect of the interaction between libraries and abstracts journals may be the increased use of the libraries fostered by the widespread dissemination of material by the abstracts journal (Tate and Wood, 1968).

Documentation Sector (DO)

If the above-mentioned annotated accession lists are compiled not from the point of view of the collection but from that of subject

---

* *Translator's note:* Key publications generally mean surveys, reviews, state-of-the-art overviews but can mean classic or seminal papers in some contexts. The term 'abstracting' is now generally preferred to 'abstracts' journal.

content (see Chapter 1, p. 7), they have thereby outgrown the library sector and developed into a current bibliography. By accepting abstracts this current bibliography finally becomes an abstracts journal. Not all abstracts journals have commenced in this way; but from general observations made by von Frauendorfer (1952), for example, it can be seen that many abstracts journals are closely linked to libraries. An example of such a development is the weekly *Landbouwdocumentatie* (Agricultural Documentation), which began as an accession list of the Ministry of Agriculture (The Hague) Library. Since this category of bibliographic material has become of paramount importance in research work, it will be expressly dealt with later. In addition to abstracts journals, current bibliographies of periodical articles without abstracts are also published; these, too, are loosely referred to as abstracts journals. Such a current bibliography is the *Bibliography of Agriculture*. Also, the indexes already mentioned (e.g. *Engineering Index*) are often called abstracts journals. The difference between the *Bibliography of Agriculture* and the current index in this field, the *Biological and Agricultural Index*, is the fact that the *Bibliography of Agriculture* sets out to cover all possible publications in the field of agriculture, while the *Index* only covers a restricted group of named periodicals.

A further type of such current title bibliographies is the series devoted to various subjects under the common title *Current Contents*. These periodicals contain only the contents lists of recent issues of periodicals. These contents lists are collected at proof stage and by use of offset reproduction very fast output is achieved.

This is also the case with the so-called KWIC (Key Word In Context) indexes, under which name current indexes to periodical contents are published in which the title of the publication is listed alphabetically under each 'important' word of the title (example: *Chemical Titles*).

Finally, in the documentation sector non-current literature lists are established, which, annotated or selective, can well provide a transition to key publications.

Research Sector (RE)

Key publications are acquiring ever greater importance. For example, a growing number of abstracts journals (e.g. *Soils and Fertilisers*, *Horticultural Abstracts*) have started to introduce, irregularly or regularly, issues with a review bibliography on some subject of topical interest. These compilations enjoy high reader appeal. In the same way, periodic reviews of the periodical literature in book form, under such titles as *Annual Review of ...*, *Progress Report of ...* or *Advances in the Field of ...* are constantly multiplying. Egan and Henkle (1956) also single out these annuals as being of special importance. In the U.S.S.R. a series has begun under the title *Itogi Nauki* (*Results of Science*) publishing review bibliographies only (Mikhailov, 1959). According to Herner (1954), an annual devoted to this 'review literature', a 'review squared' as it were, has been recommended by Adams (1953); and Wright (1953), too, expected much from this development. Perhaps these annual reviews of the literature will increasingly complement the task of the abstracts journals, indeed partly take over the task. In many fields (economics, politics, nuclear energy) an annual span is too long, but a six-monthly or quarterly publication would seem quite feasible. It would also be desirable for the review to cover in a general way the regular key publications. The principle of review publications is accepted; the problem remains of finding the best way of preserving them. One important factor is that abstracts journals mostly end up in non-commercial channels and have to fight against financial troubles, whereas surveys of the literature tend to be self-supporting. Cahn (1956) therefore advises learned societies to adopt the latter form of publication. The reason for the commercial success of surveys is that abstracts journals process and store large amounts of material, while authors of critical reviews generally process and store only a few of the publications by selection according to quality, a process which rejects the many superfluous and duplicatory publications of all sorts issued more for social reasons than for communication. These critical selections are welcome as an alternative to the personal sorting and reading of many new publications, provided that the author of the annual review article is an acknowledged authority. In future provision of literature these annual reviews will undoubtedly play an increasing role (Symposium on Critical Reviews, 1968).

TOPICALITY, COVERAGE AND ANNOTATION

If the cycle from multiplication (MU) to research sector (RE) be followed, the topicality of the auxiliary information will naturally be ever-decreasing. This means that material becomes less and less useful for current, topical information. The completeness of coverage becomes less satisfactory with increasing distance from the source. In one respect, however, a wholly essential improvement takes place: the evaluation, the annotation, becomes more important. It is completely lacking in the multiplication sector and is at its best in the research sector. This is of great importance in so far as the information is used for retrospective work. Now that abstracts journals constitute by far the most weighty aids in the documentation sector, they are worthy of further study.

ABSTRACTS JOURNALS

General Introduction

Because of the shift of emphasis in publication of the scientific literature from the book to the periodical and serials, the literature of research appears increasingly in collected form.

Scientific research is, however, not satisfied with the present form of collection. Collection, in fact, is not by subject content but is rather governed by all sorts of criteria (groupings by nationality, institutes or firms, associations or laboratories, etc.) which are of no interest to the final user. The ideal for every researcher would be one periodical which would contain the whole of the literature on his subject. Naturally, there are many highly specialised periodicals but they can never give full article coverage, even if limited to national boundaries. Even such specialised journals as *Rice Journal* and *Tea Quarterly* cannot possibly publish all the articles in these respective fields written in the world. Obviously, it is now the task of abstracting services to render this more or less useless collection of articles useful by 'de-collecting' and 're-collecting' but now in a systematic subject order, just as the user requires. There is also a form of 'abstracts journals squared', that is to say, a second collection of the same material for inquirers who are not accurately enough served in their specialised field by abstracts journals. In this way so-called 'repackaging journals' gather abstracts from other abstracts journals and reissue them reclassified, evidently in co-operation with the abstracts journals concerned.

However, it is not everywhere that abstracts journals are considered of importance by inquirers. Urquhart (1948) has established, by a questionnaire put to borrowers of scientific literature, that only a third of the references come from abstracts journals, while a third stem from other literature sources, a sixth from oral communication and a sixth from sources outside the literature. He also mentions, however, that for more recent literature, abstracts journals account for nearly half of the references and that a large proportion of those questioned were very interested in the periodical contents indexes. In the paper entitled 'The Basis of Research' (1953) the importance of the abstracts journal is regarded as diminished by the fact that the inquirer can fill in the missing pieces by means of personal contact ('Informal contacts bridge the gap'). A certain lack of enthusiasm for abstracts journals can be read between the lines in reports from Area I of the *Proceedings of the International Conference for Scientific Information, Washington* (1958). According to Pietsch (private communication), only 7 per cent of the users of *Nuclear Science Abstracts* regularly scan the abstracts, while the others wait for the appearance of the appropriate index. Another discouraging statement was made by Littleton (1969), who surveyed the use of abstracts journals in their field by agricultural economists: 23.9% *never* used the *Bibliography of Agriculture*, 41.1% never used *Journal of Economic Abstracts* and 58% *never* used their own specialised *World Economic Agricultural and Rural Sociology Abstracts*. (For more general studies on information-gathering habits, see Chapter 9.)

The difficulties facing an editorial board of an abstracts journal are problems of coverage, of scatter, of selection, of topicality and of efficiency and organisation. (For discussion of abstracts as a method of document condensation see Chapter 11, p. 106.)

Problems of Coverage

Coverage presents the most important and fundamental problem. This opinion is also shared by the users, as shown in *Figure 5*, taken from a report by Gray (1950), who questioned 1400 physicists as to the value of various qualities of abstracts journals in relation to services rendered in their work.

Coverage problems are of two kinds: those of the lacunae and those of overlapping. Problems of the first type are invariably regarded as being mainly due to: (1) the production of material in difficult languages; (2) the phenomenon of scattering, discussed separately below. The problem of the difficult languages is rather increasing with the stimulation of national activity in research and research publishing. Baker (1966) gives statistics about the steep rise of numbers of publications from, e.g., Japan, the U.S.S.R., Hungary and Poland.

First mention in any discussion of the problems of the lacunae should be given to the by now almost classic investigation by Bradford (1946a, b). It was conducted in the fields of geophysics and lubrication and showed that only about a half of the articles

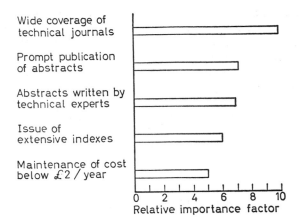

Figure 5. Relative interests of subscribers in the various qualities of an abstracts journal. (After Gray, 1950)

had been abstracted at all. Varossieau (1949) comes to the same conclusion. Jansen (1954), too, by means of a comparison of a classified catalogue with an abstracts journal reveals considerable gaps in the journal. Martyn and Slater (1964) report about 27 per cent of literature in a special field not covered at all by one of a number of abstracts journals surveyed. There are, however, more optimistic views. For example, Gorter (1954) finds that only 5 per cent of the publications tested by him in the relevant field had not been abstracted. Gorter and Mulder (1948) report that after Bradford's investigations they could establish definite improvements, especially in the fields of chemistry and physics. The literature discussed by Gorter and Mulder (1948) and, especially, by Varossieau (1949), which articles contain statistical results, is quite extensive. However, objections can be raised against the reliability of these statistics, for the collection is methodologically unsound:

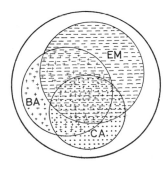

Figure 6. Coverage of 'total' research papers by abstracting services. (From Orr and Crouse, 1962, by courtesy of American Documentation)

periodical articles are counted, both those which are abstracted and those which are not. There is no indication whether the state of not being abstracted is due to critical selection by the editorial board of the abstracting journal rather than inadequate coverage.

With regard to problems of overlapping, Henkle (1951) gives examples from the fields of medicine and physics. Orr and Crouse (1962) have constructed an 'overlapping diagram' (*Figure 6*) which summarises overlapping by *Chemical Abstracts*, *Biological Abstracts* and *Excerpta Medica* in the fields of cardiovascular, endocrine and psychopharmacologic research. It is questionable whether the abstracts mention the same selection of facts and this can certainly be a matter for conjecture (see Chapter 11, p. 107). Examples of overlapping in the field of biochemistry are given by Orr (1964) and in the field of physics by Martyn (1967).

Problems of Scatter

By 'scatter' is to be understood the appearance of a certain

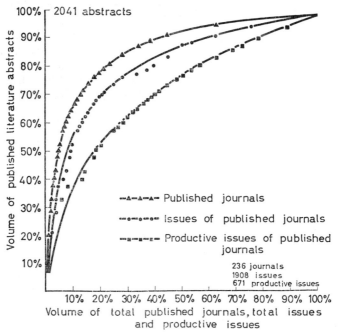

Figure 7. Representation of scatter in American periodicals. (After Taube, 1952)

percentage (varying from field to field) of articles in periodicals not specialising in that particular subject.

The range of scatter of articles on one and the same subject over many periodicals poses a great difficulty to editorial boards of abstracting journals, particularly since contributions to a certain subject can appear in a periodical in which this very article is quite unexpected. On the question of range of scatter, reference can be made to Bradford (1946a, b), Donker Duyvis (1946), Croft (1941), Bernal (1948), Grenfell (1953), Fussler (1949), Henkle (1938), Marsden (1956), Emery (1951), Taube (1952) and Varossieau (1949). This list could be considerably lengthened, as there is indeed more literature on this subject to be considered. Account may be made of articles appearing in periodicals specialising in other fields for purposes unconnected with documentation. As an example, the investigation by Gross and Woodford (1931) in the field of geology, which includes a number of similar investigations, can be quoted. The result is always the same: a major scattering of articles in all fields of knowledge. Only Emery (1951) takes the trouble to make a historical comparison, from which it can be inferred that formerly the position

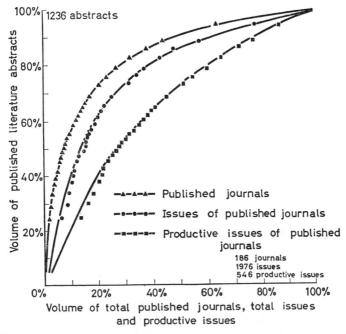

Figure 8. Representation of scatter in non-American periodicals.
(After Taube, 1952)

must have been even worse. From all this it is clear that editorial boards of periodicals keep strictly to their field, in the choice of articles, so long as they enjoy an adequate supply of contributions; at the same time, it can be seen that with a waning supply the editorial pencil is wielded much less severely. The graphic representation of range of scatter (*Figures 7 and 8*), after Taube (1952), shows the results of an investigation into 236 American periodicals (*Figure 7*), covering 1908 issues, of which 671 provided one or more abstracts (2041 in all) for the abstracts service under review. *Figure 8* shows graphically the results of an investigation into 186 non-American periodicals covering 1976 issues, of which 546 issues provided one or more abstracts (1236 in all). In both graphs the upper line is particularly important. It plots the percentage of abstracts provided (*Y*-axis) against the percentage of relevant periodicals (*X*-axis); in addition, the periodicals were arranged numerically according to the number of abstracts gleaned (from 412 abstracts to 1). *Figure 7* shows that 42.8 per cent of the American periodicals (*X*-axis) provided 91.4 per cent of the abstracts (*Y*-axis); and *Figure 8*, that 43 per cent of the non-American periodicals provided 88.5 per cent of the abstracts.

Cole (1962) introduced the concept of 'reference-scattering-coefficient' (*Figure 9*): 'The slope is clearly a measure of the extent of scatter and its numerical value, to which it is proposed to apply the term reference-scattering-coefficient, and may be a characteristic of the literature in any particular subject field.'

There is one other point which emerges from the scatter studies: there is a tendency to concentration of the really important literature in a very limited group of 'star' periodicals. Hopp (1956) refers to these periodicals as the key-periodicals. Brown (1956) finds that in chemistry 90 per cent of the references lead back to only 37 per cent of periodicals. This phenomenon of concentration will be discussed further in Chapter 9, as it may be used as an aid in searching strategy.

Scatter, as here described, arises particularly from a definite category of periodicals, the so-called 'broad' periodicals. There are in effect two categories of periodicals. The first is the so-called 'narrow' periodicals, which confine themselves to a narrowly limited field. To these belong, for example, 'house journals' of learned institutes or establishments. Examples are very specialised journals in the biological field, such as: *Mouse News, Eatonia, Selysia, Mammalian Chromosomes News Letter, Classification Programs News Letter, Drosophilia Information Service* (see Wyatt, 1967). This group includes many *series* which often also carry articles originally published in 'broad' periodicals. This collecting into a

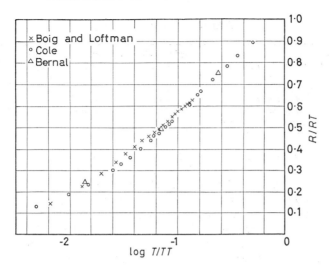

Figure 9. R/RT versus log T/TT. (For any collection of RT references derived from TT titles, R is the total number of references derived from the T most productive titles. Thus R/RT is the cumulative fraction of references and T/TT the cumulative fraction of titles. T = cumulation of titles; R = cumulation of references.) (After Cole, 1962)

smaller and more specialised context certainly helps documentation, but it is double publication, which makes the task of the 'documentation mill' all the more of a grind. This is also perhaps the reason why 'narrow' periodicals are so often missing in periodical indexes. By contrast, the second category of periodicals, the so-called 'broad' periodicals, maintain their field of interest as wide as possible, e.g. *Nature* and *Science* and publications of academies.

It was out of these 'broad' periodicals that all abstracting services arose : here the need for 'de-collecting' and 're-collecting' is greatest. If periodical literature were to deal predominantly with 'narrow' periodicals, it would be worth considering whether or not the path of 'bibliographic responsibility' could be followed, recommended by Osborn (1954) as a system of control for periodicals and serials (cf. Chapter 5). However, the 'broad' periodicals confront such a scheme with major difficulties. Who then is to assume bibliographic responsibility for these journals? Presumably this would lead to the formation of a new group of revisers and of a new abstracts journal and thereby only increase the possible overlaps and lacunae.

Wood (1956) gives an example of the variations in the various fields of science (*Table 2*).

Table 2 (After Wood, 1956)

Abstracts journal	Periodicals covered	Articles p.a.	Articles per periodical p.a.
Chemical Abstracts	6 500	80 000	12.3
Biological Abstracts	2 447	30 000	12.3
Current List Med. Lit.	1 500	109 000	72.7
Bibliography of Agriculture	22 000	100 000	4.5

It is striking how wide the scatter is in the material processed by the *Bibliography of Agriculture* and how narrow the scatter of material in *Current List of Medical Literature*.

A further and very telling example of range of scatter in the various fields of knowledge is provided by *Table 3*, drawn up by Stevens (1953) on the basis of investigations into the work of several authors. The final four columns indicate how many periodicals (in order of their productivity) contain together 25 per cent, then 50 per cent and so on, of the references; scatter is widest in the field of History of the U.S.A. Quite a different approach is adopted in *Table 4*, also by Stevens (1953). It shows the percentages of references in articles which deal with the field itself, with neighbouring fields and, finally, with different fields.

Table 3 (after Stevens, 1953)

Publication	Subject	Number of references	Number of periodicals			
			25%	50%	75%	100%
Gross and Gross (1927)	Chemistry	3 633	2	7	24	247
Fussler (1949)	Chemistry	1 085	3	5	19	131
Smith (1944)	Chem. techno-logy	21 728	3	23		
Fussler (1949)	Physics	1 279	1	3	17	134
McNeely and Crosno (1930)	Electrotechno-logy	17 991	3	9	39	
Hooker (1935)	Radiotechnology	1 506	2	8	20	
Henkle (1938)	Biochemistry	17 198	3	12	56	851
McAnally (1951)	History of U.S.A.	425	14	54	149	259

Doss (1958) studied scatter in the field of veterinary medicine and zoology and stated that in this field only 5 per cent of the publications appeared in periodicals devoted to the subject itself and 95 per cent were scattered elsewhere.

Table 4 (after Stevens, 1953)

Publication	Subject	Same field	Neighbour-ing fields	Different fields
Fussler (1949)	Chemistry	71	19	10
Fussler (1949)	Physics	63	25	12
Henkle (1938)	Biochemistry	34	55	11
Voigt (1947)	Metallurgy	61	16	23
Voigt (1947)	Mechanical engineering	27	60	13
Voigt (1947)	Soil science	14	47	39
Voigt (1947)	Dairying	39	11	50
McAnally (1951)	History of U.S.A.	31	7	62

Scatter studies (not primarily directed to the study of periodicals) based on counts of references may be incidentally useful for the study of the phenomenon of interdisciplinary scattering. Earle and Vickery (1969) found that biologists derived less from medical literature than did medical authors from biological papers.

Problems of Selection

Investigations into methods of selection have been carried out by Glass (1955), among others. He was able to establish that 30–100 per cent of the articles in the periodicals under review had actually been abstracted.

Check probes by Koster (1951) revealed that 70 per cent of the agricultural publications of the Netherlands had been abstracted in the English abstracts journals. Later Loosjes and Maltha (1952) were able to state that this 70 per cent was in fact the result of stringent selection.

The stringency of which this is an example is an indication of the endeavour to relieve the reader of the greater part of the burden of screening and thereby to dam at least in some measure the literature flood.

Screening is exceptionally responsible work. It demands a thorough training, much experience and sure judgment as to what the research worker (as information consumer so to speak) expects and what he really needs for his work. It is true that abstractors in time, and with their ranging over the literature as a whole, get a feel for the publication which really has something new or what is in demand in different quarters at the time. Despite this they must always be expected to have a personal fund of experience in the organisation of knowledge; otherwise an information service cannot measure up to the demands imposed by science and learning.

Problems of Topicality

Delays in publication of abstracts used to be frequently mentioned in the literature (e.g. Glass, 1955, on *Biological Abstracts*). Orr and Crouse (1962) have summarised in a diagram (*Figure 10*) the time lag between publication of the original and the appearance of the abstract. However, in this respect, organisation and mechanisation have brought about a good deal of improvement. Many editors of periodicals even give abstracts with the articles and not only an abstract but at times a line of descriptors.* This is called 'documentation-in-source'. It is described by, for example, Holm (1961) and Morse (1962). Editors of abstracts journals are often

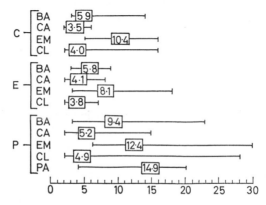

Figure 10. Interval between primary and secondary publications. The numbers represent the number of months. C=cardiovascular papers; E=endocrine papers; P=psychopharmacology papers; BA=Biological Abstracts;   CA=Chemical   Abstracts;   EM=Excerpta   Medica; PA=Psychological Abstracts; CL=Current List of Medical Literature. (From Orr and Crouse, 1962, by courtesy of American Documentation)

supplied with galley proofs before actual publication of the articles. A real advance in this field is computer-controlled type-setting. The same tape can first be used for printing the original document including its abstract (and eventual descriptors) and suitable parts of the tape can then be selected for use in the composition of the abstracts journal. (Am. Inst. of Physics Staff, 1967; Kuney, 1968; Libaur, 1969.) (See also Chapter 10, where more literature is cited.)

---

* The word descriptor will be used for any word singled out of all words of all languages for inclusion in any kind of list of terms (prescribed or free) used in a selection device as a lead to a publication or part of a publication. A descriptor is believed to be descriptive of the information of the document.

Problems of Efficiency and Organisation

The increased use of the computer has led to many developments in the production and use of abstracts journals. From editorial and publishing offices these organisations have evolved into mechanised information centres. The current-awareness function of the abstracts journal, old-style, is now supplemented by SDI (Selective Dissemination of Information) services for groups mechanically derived from the existing tapes for the production of the abstracts journals themselves (so-called repackaging journals, e.g. *Mycological Abstracts* from *Biological Abstracts*) or SDI services for personal subscribers who have handed in their 'profiles of interest' in the form of a list of descriptors. The retrospective literature-searching function of the abstracts journal is gradually being taken over by retrospective literature searches (ordered from the organisation) from the magnetic tapes available.

The editorial and publishing offices of the abstracts journals are in this way evolving into information centres and the future may eventually bring a shift from the publication activities to the information activities. This shift will include distribution of duplicate tapes by the centres to sub-centres.

As examples of prominent organisations in the field may be cited:

chemistry: CAS=Chemical Abstract Services.
biology: BIOSIS=Biosciences Information Service.
medicine: (a) Excerpta Medica (including 40 different abstracts journals).
(b) MEDLARS=Medical Literature Analysis and Retrieval Service.
physics: INSPEC=Information Service in Physics, Electrotechnology, Computers and Control.
agriculture: CAB=Commonwealth Agricultural Bureaux (including 14 abstracts journals).
nuclear sciences: INIS=International Nuclear Information System.

Co-operation between abstracts journals is promoted by ICSU (International Council of Scientific Unions) and by the American National Federation of Scientific Abstracting. A scheme for international co-operation, called UNISIST, has been launched by UNESCO.

References on these abstracting services include Houghton (1968), Ciba Symposium (1967) and special issues of *Library Trends* (1968) and *Aslib Proceedings* (1970).

Cost factors are of vital importance in the growth of the utilisation of all the possibilities offered by the use of computers in abstracting services and the corresponding information centres. These costs were the subject of the 2nd Cranfield conference, of which Harding (1970) gives a report. The proceedings of this conference have been published in the periodical *Information Storage and Retrieval*, Vol. 6 (1970), Nos. 1 and 2.

Conclusion

It can be concluded that—at least as far as the librarian and documentalist are concerned—the problem of coverage is paramount. In the case of journals it is mainly a question of coverage of periodicals and serials as such, including individual single publications. These two groups will therefore be discussed in the next chapter. The problem of 'scatter' is mainly the province of the editorial boards of the journals. These boards by responsible selection are a great help to research as a whole.

Topicality is becoming less of a problem as mechanisation enters the field of production. The cheaper the reproduction processes become, the quicker will it become customary in the future, by ordering hard copy, to consult the actual text rather than an abstract. The development of an original text delivery service, such as the *Human Relations Area File* (Guide, 1956), is already pointing the way. On the other hand, the review article is constantly growing in importance, as has already been indicated on p. 29.

For the discussion of the abstract as a condensate of the full document, see chapter 11.

References

Adams, S. (1953). *Special Libraries Association Biological Sciences Reminder* **11**, 1

American Institute of Physics Staff (1967). 'Techniques for Publication and Distribution of Information.' *A.R. Inf. Sci. Technol.* **2**, 339

ASLIB Proceedings (1970). Special Issue on Conference on International Development in Scientific Information Services. *Aslib Proc.* **22**, No. 8

Baker, D. (1966). 'Chemical Literature Expands.' *Chem. Eng. News* **44**, No. 23, 84

Basis (1953). 'The Basis of Research.' *Nature, Lond.* **171**, 1033

Bernal, J. D. (1948). 'Preliminary Analysis of Pilot Questionnaire on the Use of Scientific Literature.' In: *Report and Papers R. Soc. Scient. Inf. Conf., London.* Paper No. 46, pp. 589–637

Bradford, S. C. (1946a). 'Complete Documentation in Science and Technology.' *F.I.D. Comm.* **13**, No. 2, C1–C2

Bradford, S. C. (1946b). 'The Problem of Complete Documentation in Science and Technology.' *Proc. Brit. Soc. Int. Bibliography* **8**, No. 3, 39

Brown, C. H. (1956). *Scientific Serials (A.C.R.L. Monographs* No. 16)

Cahn, R. S. (1956). 'The Documentation of Applied Chemistry. The Future Role of Professional and Learned Societies.' *J. Docum.* **12**, No. 3, 153

Ciba Symposium (1967). *Communication in Science: Documentation and Automation.* A. de Reuck and J. Knight (eds). London

Cole, P. F. (1962). 'A New Look at Reference Scattering.' *J. Docum.* **18**, No. 2, 58

Croft, K. (1941). 'Publications and Agricultural Analysis.' *J. Chem. Educ.* **19**, No. 7, 315

Donker Duyvis, F. (1946). 'List of Periodicals for Users Specialised in Science and Technology.' *F.I.D. Comm.* **13**, No. 3, C22

Doss, M. A. (1958). 'The Importance of Peripheral Publications in the Documentation of Biology.' In: *Proc. Int. Conf. Scient. Inf., Washington.* Vol. 1, pp. 429–438

Earle, P. and Vickery, B. C. (1969). 'Subject Relations in Sciences Technology Literature.' *Aslib Proc.* **21**, No. 6, 237

Egan, M. E. and Henkle, H. H. (1956). 'Ways and Means in Which Research Workers, Executives and Others Use Information.' In: *Documentation in Action*, pp. 137–159. New York; Reinhold

Emery, K. O. (1951). 'Trends in Literature of Sedimentology.' *J. Sed. Petrol.* **21**, No. 2, 105

Frauendorfer, S. von (1952). 'Dokumentationsfragen der Land- und Forstwirtschaft, der Holzforschung und des Ernährungswesens.' *Nachr. Dokum.* **3**, 4

Fussler, H. H. (1949). 'Characteristics of Research Literature Used by Chemists and Physicists in the U.S.A.' *Libr. Q.* **19**, 19, 119

Glass, B. (1955). 'Survey of Biological Abstracting.' *Science, N.Y.* **121**, 31–47, 583

Gorter, A. (1954). 'Literatuuronderzoek in kleine en middelgrote Bedrijven en Instellingen.' In: *Symposium over Literatuuronderzoek.* pp. 18–26. The Hague; NIDER Publ. No. 9, 2nd series

Gorter, A. and Mulder, S. H. (1948). 'Referaten als Hulpmiddel bij de Literatuur documentatie.' *Bibliotheekleven* **33**, 237

Gray, D. E. (1950). *Study of Physics Abstracting: Final Report.* New York; American Institute of Physics

Grenfell, D. (1953). *Periodicals and Serials. Their Treatment in Special Libraries.* London; ASLIB (2nd edn, 1965)

Gross, P. L. K. and Woodford, A. O. (1931). 'Serial Literature Used by American Geologists.' *Science, N.Y.* **73**, 660

Guide (1956). *Guide to the Use of the Files.* New Haven (Conn.), Human Relation Area Files Inc.

Harding, P. (1970). 'Second Cranfield International Conference on Information Storage and Retrieval.' *Program* **4**, No. 1, 42

Henkle, H. H. (1938). 'The Periodical Literature of Biochemistry.' *Bull. A.L.A.* **27**, 139

Henkle, H. H. (1951). 'The Natural Sciences: Characteristics of Literature, Problem of Use and Bibliographical Organization in the Field.' In: *Bibliographical Organization.* pp. 140–160. Chicago; Univ. of Chicago Press

Herner, S. (1954). 'Information Gathering Habits of Workers in Pure and Applied Sciences.' *Ind. Engng Chem.* **46**, 229

Holm, B. E. (1961). 'Information Retrieval. A Solution.' *Chem. Engng Progr.* **57**, No. 6, 73

Hopp, R. H. (1956). 'A Study of the Problem of Complete Documentation in Science and Technology.' *Diss. Abstr.* **16**, 1689

Houghton, B. (ed.) (1968). *Computer-based Information Retrieval Systems.* London; Bingley

Jansen, H. (1954). *Enige Aspecten van de systematische Catalogus van de I.V.P. en S.V.P.* Bibliotheek en een Vergelijking met Plant Breeding Abstracts als Literatuurbron. Stencilled report. Wageningen; Institute of Plant Breeding

Koster, G. (1951). *In hoeverre Worden Nederlandse Publicaties op het gebied van de Landbouw in buitenlandse Referaattijdschriften Gerefereed?* Stencilled report. Wageningen; Centrum voor Landbouwdocumentatie

Kuney, J. H. (1968). 'Publication and Distribution of Information.' *A.R. Inf. Sci. Technol.* **3**, 31

Libaur, F. B. (1969). 'A New Generalized Model for Information Transfer. A Systems Approach.' *Am. Docum.* **20**, 381

Library Trends (1968). Special Issue on Science Abstracts Services (ed. F. E. Mohrhardt). *Library Trends* **16**, No. 3 (January)

Littleton, I. T. (1969). 'The Literature of Agricultural Economics.' *Libr. Q.* **39**, No. 2, 140

Loosjes, T. P. and Maltha, D. J. (1952). *Landbouwdocumentatie in England.* Rapport over het Toegankelijk maken van de Resultaten van Landbouwkundig Onderzoek. The Hague, 1954; Documentation Section, Min. of Agric.

Marsden, A. W. (1956). 'Documentation of Applied Chemistry.' In: *1st Int. Congr. Chemy Ind.* (14th Jan.), 36–45

Martyn, J. (1967). 'Tests on Abstracts Journals. Coverage, Overlap and Indexing.' *J. Docum.* **23**, No. 1, 45

Martyn, J. and Slater, M. (1964). 'Tests on Abstracts Journals.' *J. Docum.* **20**, No. 4, 212

Mikhailov, A. L. (1959). 'Aims and Purposes of Scientific Information.' *UNESCO Bull. Libr.* **13**, 262

Morse, R. D. (1962). 'A.I.Ch.E. Information Retrieval Activities.' *Am. Docum.* **13**, No. 1, 69

Orr, R. H. (1964). 'Communication Problems in Biochemistry Research: Report of a Study.' *Fed. Proc.* **23**, No. 5, 117

Orr, R. H. and Crouse, E. M. (1962). 'Secondary Publications in Cardiovascular, Endocrine and Psychopharmacologic Research.' *Am. Docum.* **13**, No. 2, 197

Osborn, A. D. (1954). 'The Future of the Union List of Serials.' *College Res. Libr.* **15**, No. 1, 118

Quartz, B. M. (1953). 'Policies for Analysing Monograph Series, Pt. 1 College Libr.' *Serial Slants*, **4**, No. 3, 124

Rogers, F. B. (1964). 'The Relation of Library Catalogs to Abstracting and Indexing Services.' *Libr. Q.* **34**, No. 1, 106

Stevens, R. E. (1953). 'Characteristics of Subject Literature.' In: *ACRL Monographs* No. 6, 10

Symposium on Critical Reviews (1968). *J. Chem. Doc.* **8**, 231

Tate, F. A. and Wood, J. L. (1968). 'Libraries and Abstracting and Indexing Services. A Study of Interdependency.' *Libr. Trends* **16**, No. 3, 353

Taube, M. (1952). 'Possibilities for Cooperative Work in Subject Controls.' *Am. Docum.* **3**, No. 1, 21

Urquhart, D. J. (1948). 'The Organization of the Distribution of Scientific and Technical Information.' In: *Report and Papers R. Soc. Scient. Inf. Conf., London.* pp. 75–89

Varossieau, W. W. (1949). *Een Onderzoek aangaande Referaatdiensten op het Gebied der zuivere en toegepaste Natuurwetenschappen.* The Hague; *NIDER* Publ. No. 285.

Wood, G. C. (1956). 'Biological Subject Indexing and Information Retrieval by Means of Punched Cards.' *Spec. Libr.* **47**, No. 1, 26

Wright. W. E. (1953). 'The Subject Approach to Knowledge: Historical Aspects and Purposes.' In: Tauber, M. F. (ed.). *The Subject Analysis of Library Materials.* pp. 8–15. New York; School of Library Service, Columbia Univ.

Wyatt, H. V. (1967). 'Research Newsletters in the Biological Sciences.' *J. Docum.* **23**, 321

FIVE

BIBLIOGRAPHIC CONTROL OF PERIODICALS AND SERIALS AS SUCH

INTRODUCTION

There are some further points to be raised in regard to the biblio-graphic control of periodicals and serials by editorial boards of abstracting journals. Glass (1955) shows in his conspectus of *Biological Abstracts* that in 1947–9 they covered 10 per cent of the current biological and partly biological research and review journals. Two further conclusions can be gathered from two investigations which were carried out independently of each other in the field of medicine and which came to astonishingly similar results: Hyslop (1953) investigated 2674 medical periodicals and found that, of these, 43 per cent were ignored by the 34 abstracting services; Larkey (1952) established in the case of 4695 medical periodicals that the five most important abstracts journals (*Biological Abstracts, Chemical Abstracts, Current List, Index Medicus, Excerpta Medica*) did not cover 50 per cent of these periodicals.

Just how much effort is required by an editorial board of an abstracting journal to obtain a true general picture of the periodicals and serials carrying material in their field is shown by, for example, Gray (1950) in the field of physics.

The major lacuna in the circle *Figure 4* (p. 28) results from the fact that editors of abstracting journals cannot possibly evaluate all stocks and accessions of all the libraries in the world. An attempt certainly was made by, for example, Crane (1953), who prepared an exact tabulation of the periodicals covered in each country by *Chemical Abstracts*. A useful aid could also be such compilations as that of Gummer (1956), who worked out a synopsis of all lists of periodicals published in England.

DESCRIPTION OF THE VARIOUS TYPES OF LISTS OF PERIODICALS

Lists of periodicals could be classified by place of publication, subject matter, method of use or utility. The criteria chosen here for the work are compiler and motive. In choosing the compiler, reference is made to the four sectors of activity, which have already been considered, namely multiplication, librarianship, documentation and research.

Multiplication Sector

Besides national editions, of national publishers' organisations, there is one international annotated bibliography of national bibliographies of periodicals published by IFLA (Duprat, Liutova and Bossuat, 1969).

Library Sector

International lists are to be found with acquisition as the *raison d'être* (example: *Ulrich's International Periodicals Directory*, 13th edn, New York, 1969) and also lists limited to periodicals appearing within a strictly limited geographical area (example: Toase. *Guide to Current British Periodicals*. London, 1962). Stimulus towards localisation is provided by published and unpublished catalogues of periodical holdings limited to *one* library; similarly with catalogues of periodical holdings of a whole country (as they are found in a geographical catalogue of periodicals).

Location material which refers to more than one library includes all *union* catalogues and inventory union lists, which are mostly regional (example: *British Union–Catalogue of Periodicals*. 4 volumes and supplements). There are also *inventory union lists* with material on a particular country—for example, *Verzeichnis schweizerischer Zeitschriften in den wissenschaftlichen Bibliotheken West-Deutschlands*, Berlin 1950. (Union list of Swiss periodicals in research libraries of West Germany, Berlin, 1950.) Under the same heading belong slip catalogues, called 'Union Catalogues'. Brummel (1951, 1952, 1956) has studied the whole complex of union catalogues, their installation, maintenance and use, and other unsolved problems. There are bibliographies devoted to this material, e.g. those by Mummendey (1939), Haskell and Brown (1943) and

Downs (1942). Finally, there is even an example of an international union list of periodicals in geography by Harris and Fellmann (1950) covering two English, four French and 70 American and Canadian libraries, further details of which are given by Gummer (1956).

Documentation Sector

Here, in relation to bibliographic control, attention is first paid to lists published by editorial boards of abstracting journals (example: list of periodicals published by *Chemical Abstracts*). In the national field there are compilations commissioned by UNESCO. For details of the bibliographic activity deployed by UNESCO reference should be made to the essay by Carter (1954). In the Netherlands there is the well-known *List of Scientific and Learned Periodicals in the Netherlands* (1953) and in Sweden the list by Düring. In addition, without direct promotion by UNESCO, there are such national compilations as, for example, the Belgian list by van Hove (1951; and supplement, 1955). Most of the others are bookseller publications, such as the Danish list by Bredsdorff (1949) or the list *Deutsche Zeitschriften 1945–1949* (1950). As an example of the need for title checking resulting in a compilation, *World Medical Periodicals*, 3rd edn (1961) can be cited. Check lists enable the greatest possible degree of standardisation in abbreviations to be achieved. Important literature on the subject of *periodica abbreviata* is to be found in the work of Kent (1954), Bishop (1953) and Reid (1954). Kent deals with the historical aspect, while Reid and Bishop propose new methods.

Research Sector

There are numerous international lists in the research sector (which are mostly titled 'bibliography'; this word is to be avoided in this context, as it might be wrongly taken to mean controls. Examples are: Claassen, *Entomological Periodicals* (1945); Frykholm, *Apiculture* (1954); Guilletmot, *Dairying* (1953); Roq, *Petroleum* (1945). To this genre belong all the so-called literature guides, which mostly devote a chapter to periodicals: some examples are Whitford (1954), Bottle (1962) and Smith (1962). Other examples are the *American Potato Yearbook*, which publishes regularly a list of periodicals on the potato, and Verdoorn (1945), who has compiled a list of South American botanical periodicals.

Other lists refer to material in a country. As examples in the

Netherlands, Verkerk (1954) and Doorenbos (1951) can be quoted. The lists in this sector are all aimed at helping in literature searches in some field of knowledge, but this does not always mean that they are compiled by the research workers themselves.

It can be seen that lists of periodicals are compiled for many different motives and that all these types are useful, in varying degrees, for bibliographic control, even though the main aim of the publisher was to provide a location check list.

What then, in fact, are the actual prospects of full coverage?

THE QUANTITY OF PERIODICALS COVERED

Some statistical material can help in providing an answer to this question. Stewart (1953) ascertained that the *British Union Catalogue of Periodicals* covers 40 per cent of titles which are missing in the American union list. Donker Duyvis (1946) mentions that of the 840 electrotechnical periodicals in the world 41 per cent are contained in the *World List** and 58 per cent in the list compiled by NIDER. Furthermore, the *World List* has 27 000 titles not in NIDER but the latter contains another 8000 titles not in the *World List*. In a comparison of *World Medical Periodicals* (3483 titles) with the list of the *Welch Medical Library Projects* (4454 titles) only 2211 titles were found to be common to both major compilations (International Advisory Committee for Documentation and Terminology in Pure and Applied Sciences, 1955). A personal investigation by the author showed that *New Serial Titles*, which sets out to cover new periodicals from 1950 to 1953, listed only one-third of the agricultural periodicals of the Netherlands, which had been heralded as new arrivals for this period in the first edition of the list of scientific and learned periodicals in the Netherlands (1953) mentioned on p. 48.

CURRENT BIBLIOGRAPHIES OF PERIODICALS

An important requirement of all lists of periodicals is that they be up-to-date. In the various sectors some attempt is made to meet this requirement. In the multiplication sector, for example, reference can be made to *Faxons Bulletin of Bibliography*, *Stechert Hafner News Letter* and the 'Recent Titular Changes and Amalgamations'

---

* Applies to 3rd edn. 4th edn, published 1965, is more complete.

noted in *Willing's Press Guide*. In the library sector mention can be made of the 'New Periodicals' listed in the library journal *College and Research Libraries* and of the accessions lists of libraries. In the documentation sector examples of updating can be found under 'Magazine Notes' in the *Industrial Arts Index*, under 'New Periodicals' in *Biological Abstracts* and in the *Quarterly Bulletin of the International Association of Agricultural Librarians and Documentalists*.

It is tempting to equate periodicals with people. Lists of periodicals become obsolete as quickly as address books. A periodical is said to have a biography and this term has some anthropomorphic characteristics. For example, the headings in the periodical *Vital Notes* (1952), published by the Medical Library Association, cover, as it were, the personalia of the world of medical periodicals: births, deaths, marriages, divorces, remarriages, change of name, and even hibernations and expectations. The biography of a periodical can be as thrilling reading as the biography of a person. There is another similarity between periodicals and people: the coincidence of peak birthrate in the post-war period (with peak infant mortality in periodicals only).

Family relationships are also of interest: for example, *Archives Néerlandaises des Sciences Exactes et Naturelles* is the great-grandmother of five extant research journals in the Nederlands. What makes for difficulties in the case of periodicals is that there is no central secretariat for the control of periodicals.

A special development in the U.S.A. must be considered which attempts to keep up with current biographical details of periodicals. As an addition to the *Union List of Serials* and its two supplements there appeared *Serial Titles Newly Received*, which reports periodicals newly accessioned to the Library of Congress as well as those newly acquired by the New York Public Library, which at the same time cancelled its bulletin entry for new periodicals. Since then, this publication has metamorphosed into a periodical *New Serial Titles*, which also covers material *not* carried by the libraries concerned; see also Cole (1953). This change means that a compilation originally designed for local purposes only becomes a tool for bibliographic control. A somewhat similar development is seen in the Netherlands: if the C.C.P. (Union-Catalogue of Periodicals) is unable to satisfy a request, the periodicals missing are noted and acquired by *one* of the big scientific libraries after consultation among themselves. Osborn (1954) describes how in the U.S.A. two libraries undertake bibliographic responsibility for a certain section, i.e. that they are bound to maintain subscriptions and report discontinuations; they must also notify all changes concerning the periodical. Thus, as Osborn states, 'follow-up work, an area in which libraries have generally been notoriously weak, is thereby guaranteed'.

References

Bishop, Ch. (1953). 'An Integrated Approach to Documentation Problems.' *Am. Docum.* **4**, No. 4, 54
Bottle, R. T. (Ed.) (1969). *The Use of Chemical Literature.* 2nd edn. London; Butterworths
Bredsdorff, V. (1949). *Dansk Tidskriftfortegnelse.* 1949. Copenhagen
Brummel, L. (1951). 'Union Catalogues.' *Libri* **1**, No. 3, 201
Brummel, L. (1952). 'Some Problems of Union Catalogues.' *Actes Comm. Int. Biblth.* **17**, 88
Brummel, L. (1956). *Union Catalogues. Their Problems and Organization* (UNESCO Bibliographical Handbook No. 6). Paris; UNESCO
Carter, E. J. (1954). 'UNESCO's Bibliographical Programme.' *Am. Docum.* **5**, No. 1, 1
Claassen, E. S. (1945). 'Annotated Checklist of Entomological Periodicals.' *Ann. Ent. Soc. Am.* **38**, No. 3, 403
Cole, B. (1953). 'Union of Serials Interim Report.' *Spec. Libr.* **44**, No. 6, 243
Crane, E. J. (1953). 'National Sources of Scientific Journals.' *Chem. Engng News* **31**, 864
Donker Duyvis, F. (1946). 'List of Periodicals for Users Specialised in Science and Technology.' *F.I.D. Comm.* **13**, No. 3, C22
Doorenbos, J. (1951). 'Nederlandse Tijdschriften op Tuinbouwgebied.' *Meded. Dir. Tuinb.* **14**, 372
Downs, R. B. (1942). *Union Catalogues in the United States.* Chicago; A.L.A.
Duprat, G., Liutova, K. and Bossuat, M. L. (1969). *Bibliographie des Repertoires Nationaux de Périodiques en cours* Paris; UNESCO (Manuels Bibliographiques de l'UNESCO No. 12)
Frykholm, L. (1954). *Förteckning över Bitidskrifter.* Uppsala
Glass, B. (1955). 'Survey of Biological Abstracting.' *Science, N.Y.* **121**, 583
Gray, D. E. (1950). *Study of Physics Abstracting. Final Report.* New York; American Institute of Physics
Guilletmot, A. (1953). 'Förteckning över periodiska Publikationer rörande Mejerihanteringen.' *Nord. JordbrForsk.* **35**, No. 2, 115
Gummer, H. M. (1956). 'Catalogues and Bibliographies of Periodicals.' *J. Docum.* **12**, No. 1, 24
Harris, C. D. and Fellmann, J. D. (1950). *A Union List of Geographical Serials* (2nd edn). Chicago; Univ. of Chicago Dept. of Geography. Research Paper 10
Haskell, D. C. and Brown, K. (1943). 'Bibliography of Union Lists of Serials.' In: *Union List of Serials* (2nd edn). pp. 3053–3065. New York
Hove, J. van (1951 and 1955). *Repertorium van de in België verschijnende Tijdschriften.* Brussels
Hyslop, M. R. (1953). 'Documentalists Consider Machine Techniques.' *Spec. Libr.* **44**, No. 5, 196
International Advisory Committee for Documentation and Terminology

in Pure and Applied Sciences (1955). *Final Report. 2nd meeting, London, 1955.* UNESCO NS/135 Paris, 13.4.56

Kent, F. L. (1954). 'Periodica Abbreviata and International Standardization.' *J. Docum.* **10**, No. 2, 59

Larkey, S. V. (1952). 'Some Approaches to the Problem of Indexing.' *Bull. Med. Libr. Ass.* **40**, No. 2, 107

List (1950). *Deutsche Zeitschriften, 1945–1949.* Frankfurt/M.; Buchhändler Vereingung

List (1953). *List of Scientific and Learned Periodicals in the Netherlands.* The Hague

Mummendey, R. (1939). *Bibliographie der Gesamt-Zeitschriftenverzeichnisse.* Cologne (Kölner Bibliographische Arbeiten Bd. 4)

Osborn, A. D. (1954). 'The Future of the Union List of Serials.' *College Res. Libr.* **15**, No. 1, 26, 118

Reid, J. B. (1954). 'Chronological Sigils.' *Am. Docum.* **5**, No. 1, 26

Roq. M. M. *et al.* (1945). 'Petroleum Periodicals.' *Spec. Libr.* **36**, No. 8, 376

Smith, R. C. (1962). *Guide to the Literature of the Zoological Sciences* (6th edn). Minneapolis; Burgess Publ. Co.

Stewart, J. D. (1953). 'The British Union Catalogue of Periodicals.' *Libr. Ass. Rec.* **55**, 248

Verdoorn, F. (Ed.) (1945). *Plants and Plant Science in Latin America.* Waltham; Publ. Chronica Botanica

Verkerk, P. (1954). *Per Provincie geordende Titels van Jaarverslagen over akker—en weidebouwkundig onderzoek van Landbouwleraren. Rijkslandbouwconsulten, Verenigingen voor Bedrijfsvoorlichting, Proefboerderijen.* Wageningen; Central Instituut v. Landbouwkundig Underzoek

Whitford, R. H. (1954). *Physics Literature.* New Brunswick; Scarecrow Press

SIX

BIBLIOGRAPHIC CONTROL OF SINGLE PUBLICATIONS

INTRODUCTION

Gorter and Mulder (1948) have described the collecting and control of single publications as the major problem facing editors of abstracting journals; the same comment was made by Marsden (1956) with reference to his journal *Dairy Science Abstracts*. Of the material abstracted, 30 per cent consists of single publications (bulletins, patents and irregular material). Technical reports in this respect already pose a problem peculiar to themselves.

Varossieau (1949) concludes that only 5 per cent of the abstracting services do not abstract articles in periodicals but that 30, 45 and 59 per cent do not deal with books, dissertations and patents, respectively. It is clear that the countercurrents in the development of the shift from book to periodical render bibliographic control of single publications still more difficult. Suggestions for the splitting up of periodicals naturally find little support among those responsible for bibliographic control. With regard to such splitting as has already been carried out, the only solution is to arrange the publications in question into series as soon as possible so that the material can at least be found. The remarkable fact remains that the 'imposition of condition of sale' of publications, from which the necessity of de-collection and re-collection first arose, should be welcome from the point of view of bibliographic control, while it is this very lumping together of material that is the cause of those difficulties, as described on p. 31.

BIBLIOGRAPHIC MATERIAL

MU Sector

In the MU sector national bibliographies form the most important material. Besides these, there are publishers' catalogues (particularly important in the case of specialised publishers) and joint publishers' catalogues (Germany, Switzerland). How far does national bibliography meet modern demands, and to what extent does it fail? In answer to this question, Clapp (1951) cites bodies in various countries concerned with national bibliography. Wadsworth (1954) has drawn attention to the fact that the national publications are, in general, prone to gaps and much too uneven in design and execution. Larsen (1953) has given a conspectus of the state of national bibliographies, based on the UNESCO recommendations. The extent of the spate of publications outside the book trade is shown by Glass (1955), who reports that in 1954 the *Deutsche Nationalbibliographie*, Series B, non-book trade publications, covered 18 000 items, i.e. two-fifths of the total national production. It is apparent that mechanisation (of

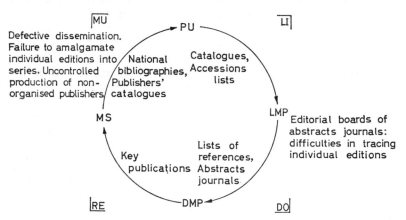

Figure 11. Contributions of bibliographic material from the various sectors to bibliographical control of individual publications and the lacunae in this field

which *British National Bibliography* and *Deutsche Nationalbibliographie* are examples) will help a great deal with the difficulties mentioned.

LI Sector

Catalogues are important in the LI sector, especially classified catalogues in so far as they exist. Internationally shared cataloguing and the sponsoring of such projects as that of the Library of Congress

(MARC: Machine-Readable Cataloguing) will increasingly improve the situation in this sector.

DO and RE Sector

The DO and RE sectors are not interested in the form in which anything is published; neither do they take any particularly bibliographic control measures for publications not contained in periodicals or series. This type of publication is dealt with in the same manner as articles in periodicals; the bibliographic material (DO sector—abstracts, periodical and literature indexes; RE sector —monographs, key publications) is in effect the same as that mentioned in the discussion of periodical articles.

Lacunae

Figure 11 shows where gaps occur: (1) The productions of non-organised publishers, printers and duplicating businesses are not controlled. (2) Little care is bestowed upon editions which could easily have appeared as series. (3) Issuing bodies do not do enough in the way of dissemination. Already items 1–3 cause a lacuna between MU and LI. (4) Editorial boards of abstracts journals cannot cover all libraries and their collections.

References

Clapp, V. W. (1951). 'The Role of Bibliographical Organization in Contemporary Civilization.' In: *Bibliographical Organization.* pp. 3–23. Chicago; Univ. of Chicago Press

Glass, A.-M. (1955). 'Die bibliographische Erschliessung der Schriften, die nicht im Buchhandel erscheinen.' *Dokumentation* 2, No. 2, 26–30

Gorter, A. and Mulder, S. H. (1948). 'Referaten als hulpmiddel bij de literatuur documentatie.' *Bibliotheekleven* 33, 237

Larsen, K. (1953). 'National Bibliographical Services, Their Creation and Operation.' Prepared in Accordance with the Recommendation of the Int. Adv. Comm. Biblphy. Paris; UNESCO

Marsden, A. W. (1956). 'Documentation of Applied Chemistry.' *1st Int. Congr. Chemy Ind. 14th Jan. 1956.* pp. 36–45

Varossieau, W. W. (1949). *Een onderzoek aangaande referaatdiensten op het gebied der zuivere en toegepaste natuurwetenschappen.* The Hague; NIDER Publ. No. 285

Wadsworth, R. W. (1954). 'Some Lacunae in Foreign Bibliographies.' *Libr. Q.* 24, No. 2, 124

SEVEN

SOLUTIONS TO PROBLEMS OF BIBLIOGRAPHIC CONTROL

INTRODUCTION

Proposals for a reduction in publication will be discussed first and then proposals for an improvement in national bibliographic control (see *Figure 12*).

PROPOSALS FOR REDUCTION IN PUBLICATION

Limitation of Publications

Weiss (1945) and Laclémandière (1955) wish to limit the number of publications by means of preprints, which would then be screened. Weiss (1945) proposes cheap reproduction methods and that only publications reaching a certain level of maturity should be allowed into print. Then it would no longer be a matter of counting pages (!) as a measure of a research worker's output. It is conceivable that publications could be divided into those concerned with establishing priority and those detailing completed work. The first category could evolve into something like patent literature; the second, into official and prestige literature (see also Chapter 2, where certain items in this field have already been mentioned). Brimble (1955) calls for stricter control by editorial boards of periodicals and states that they should demand condensation of the articles submitted. Cleland (1956) wishes to publish abstracts only and to supply the originals on demand in microform. De Grolier (1959) reports on the results of a questionnaire to two hundred scientists, from which a state of 'over-communication' can be deduced.

Proposals Stressing the Use of Periodicals and Series

Recommendations stressing the use of periodicals and serials are made by Sale (1954) and Cleverdon (1953). If the librarian were to be brought into early consultation on the publication policy of his

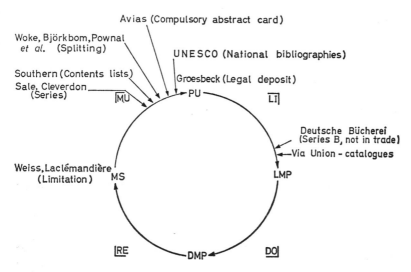

Figure 12. The various suggestions for solving problems of bibliographic control

institute or establishment, he would see to it that series title, number, addresses for subscription and publisher and such matters were considered. There is also a counter-proposal to this (Woke, 1951, and others) which has already been described (p. 14). The proposals of Southern (1953) and Avias (1952) likewise belong to this group.

PROPOSALS FOR MEANS OF IMPROVING NATIONAL BIBLIOGRAPHIES

UNESCO supports attempts to arrive at a solution by improving national bibliographies.

With Deposit Copy (Statutory copy)

Groesbeck (1951) and Malclès (1950) have high hopes of this method, but one is virtually helpless in view of all the small reproduction centres (including, in fact, every stencil machine).

Without Deposit Copy

Here the *sine qua non* is an improvement in bibliographic control—for example, by an extremely active library such as the Deutsche Bücherei. In most countries, however, one finds no trace of this improvement and one limits oneself to the control offered by known publishers.

A PERSONAL PROPOSAL

As has already been shown in dealing with the bibliographic control of periodicals and serials as such, much is expected from close co-operation with existing union catalogue organisations in this respect. Larsen (1953) has already proposed that, in building up a national bibliography system, the union catalogue should not be the second but rather the first priority. Brummel (1956) recommends close liaison between national bibliography and union catalogue and stresses the potential of a retrospective national bibliography, provided that national and foreign material are kept apart in the union catalogue. Egger (1956) does not admit the existence of subsidiary functions for a union catalogue and quotes literature on the subject.

The author's proposal could be formulated as follows: For periodicals and serials as such, as well as for single publications, the use of the national part of the national union catalogue should be considered a means of bibliographic control. In this manner a nearly 'watertight' bibliographic control is achieved, since each publication should be represented in at least one of the major libraries covered by the outlier scheme of the union catalogue concerned. If the organisation is optimalised, the setting-up of bibliographic control follows as a by-product. In many countries this union catalogue system is already in existence, so that no new set-up is necessary. The disadvantages, or rather the conditions which should be fulfilled but which are often only partially met or not met at all, are:

1. The necessity of dividing the union catalogue into national and international sections.
2. The time lag between appearance of a publication and its entry in the union catalogue.
3. The participation of sufficient representative libraries in all fields.

Observations on (1)

In many cases the division would result in work economy in the daily use of the union catalogue itself. It presupposes that the national section would be separated. This part of the catalogue would, however, be less consulted. A separation of the national and international sections would be to the advantage of the latter, more intensively used, part. Just such a division has been carried out in Switzerland between the national union catalogue and the non-Swiss publications. The catalogue of the Swiss National Library is automatically the first catalogue; the second, that is the international, is arranged separately (Egger, 1956).

Observations on (2)

The time lag between the appearance of a publication and its entry in the union catalogue depends, in the case of important works which are acquired by many libraries, only on the speed of the quickest library, which is naturally favourable. The fewer libraries that acquire a title, the more on average will its cataloguing be delayed.

Observations on (3)

These demands are met wherever the organisation of a union catalogue is reasonably well developed. In each country the investigation of possibilities can mean the first step towards a solution; perhaps the Netherlands could set an example if a start were made with periodicals and serials. In any case, where the union catalogue most fully meets requirements, there should the first experiment be made in the combination of union catalogue and bibliography.

CONCLUSION

Finally, it may be mentioned that certain quarters are of the opinion that the whole method of publication will be so fundamentally altered that the above-mentioned problems will no longer exist. The answer of the research worker to the increasing flow of literature is discussed in Chapter 9.

References

Avias, J. (1952). 'International Organization of Scientific Documentation Based on Legislation.' *Science, N.Y.* **115**, 250

Brimble, L. J. F. (1955). 'Rushing into Print.' *Science, N.Y.* **122**, 18

Brummel, L. (1956). *'Union Catalogues. Their Problems and Organization.'* Paris; UNESCO

Cleland, R. E. (1956). 'Analysis of Trends in Biological Literature. Plant Science.' *Biol. Abstr.* **30**, No. 24, 59

Cleverdon, C. (1953). 'The Documentalist.' *ASLIB Aeronautical Group Newsletter* Nov. 1953, 10

Egger, E. (1956). 'Gesamtkatalog, Aufbau und Organization eines Gesamt-katalogs in Hinblick auf die Benutzung.' *Libri* **6**, No. 2, 97

Groesbeck, J. A. (1951). 'Discovering Sources of Research Material.' *Am. Docum.* **2**, No. 2, 80

Grolier, E. de (1959). 'Summary of the Discussions of Area 1.' *Proc. Int. Conf. Scient. Inf., Washington.* Vol. 1, p. 311

Laclémandière, J. de (1955). 'Les publicationes périodiques répondent-elles aux besoins documentaires.' *ABCD* **2**, 51

Larsen, K. (1953). *National Bibliographic Services, their Creation and Operation.* Prepared in accordance with the recommendation of the Int. Adv. Comm. Biblphy. Paris; UNESCO

Malclès, L. N. (1950). *Les sources du travail bibliographique.* Genève; E. Droz

Sale, R. C. (1954). 'The Function of Technical Reports.' *ASLIB Proc.* **6**, No. 4, 268

Southern, W. A. (1953). 'Library Bulletins in an American Industrial Library.' *ASLIB Proc.* **5**, No. 4, 320

Weiss, P. (1945). 'Biological Research. Strategy and Publication Policy.' *Science, N.Y.* **101**, 101

Woke, P. A. (1951). 'Considerations on Utilization of Scientific Literature.' *Science, N.Y.* **113**, 399

EIGHT

THE RESEARCH WORKER AND DOCUMENTATION OF THE LITERATURE

INTRODUCTION

After the discussion of active documentation attention is now turned to passive documentation. Here the whole complex of questions which concern the researcher and his own, or delegated, literature search is considered.

Even though literature really forms part of scientific research, we shall see that in some groups of research workers there is a tendency to delegate a part of the documentation of the literature. This has led, often within the framework of libraries or with their co-operation, to the formation of screening and documentation services and to the rise of a new vocation: that of the documentalist. In this connection Fairthorne (1969) has written: '... a delegate not only must be able to do what he is told, but must also be able to be told what to do'.

The five phases of a complete retrospective literature survey are:

1. Definition and limitation of subject matter.
2. Search strategy scheme.
3. Looking up titles.
4. Retrieval of the literature itself.
5. Evaluation.

Phase 1 is the research worker's own field. Phase 4 belongs to the library serving the researcher concerned. Phase 5, again, is peculiar to the research worker. In the following discussion the term 'literature research' refers to phases (2) and (3) of the retrospective survey.

In this presentation the material is divided into four stages. In stage *A* the documentalist has as yet no function between researcher and librarian; but he has in stage *B*. Documentation plays the most important role in stage *C*. In stage *D* the documentation services are serving other 'clients', no longer the research worker; however, to complete the picture, this stage *D* must be mentioned. The whole picture is that of an increasing tendency to delegate the work of literature searching.

STAGE-*A*—A STUDY OF THE LITERATURE IN A SPECIAL LIBRARY

(a) From the historic point of view this stage is the original form of literature research. Nothing as yet separates the scholar from the material; in most cases he can work in a reference library, where the catalogue, as a guide to the stock, scarcely needs to play any role. Bibliographic material, too, is still worked up by the scholar himself. The only obstacles between the scholar and the material are the selection of entry words in the title indexes and perhaps the use of a classification scheme.

(b) The bibliographic material consists of bibliographies. As soon as this material becomes voluminous enough, the first bibliography of bibliographies will appear (1664: Labbé, *Bibliotheca bibliothecarum*). There is no talk yet of documentation. The library catalogue is here—bibliographically speaking—a tower of strength, because many more individual works (books and pamphlets) appear than periodicals and serials, which, at least in the larger libraries, are not analytically catalogued by article. Accession lists of periodical articles play as yet a comparatively modest role.

(c) This situation can still be found today in the university faculties of the humanities. Its characteristic, from the documentation point of view, is the lack of competing sources of knowledge: the literature is the most important object of study and the scholar is so at home in it that there is no question at all of work delegation; documentation services do not play a part. These scholars browse only too gladly in periodicals; perhaps books other than the one immediately sought can be of help to them either methodologically or otherwise, now or in the foreseeable future (compare, for example, Eppelsheimer, 1951). Evans (1956) even thinks that in the humanities it could be useful to work through the material a second time and study it again; to him the task of documentation in these scholarly disciplines appears not to be of such importance as in the more experimental ones, where the repetition of the same experiment would be looked upon as waste, unless intended as a check.

Characteristic of the outlook of the humanities on documentation is the following declaration (after van Alphen, 1953):

People unfortunately do not sufficiently recognise that so-called documentation, in the meaning of provision of references on a certain subject, in which direction libraries and other institutions are being ever more urged by the public, this documentation harbours great dangers for learning.... If the librarian does not want to mislead people, he would do well to avoid this type of 'documentation', indeed he must shun it altogether. The one and only thing he can and should do with a clear conscience is to show the scholar the way within the collection of the library.

The antithesis of this declaration can be seen in a successful delegation of literature search in the U.S.A. in the field of the humanities, which will be dealt with below.

(d) Instruction. This is given to the students by the professor or under his aegis.

STAGE B—EXPERIMENT IN CONJUNCTION WITH STUDY OF THE LITERATURE

(a) Historically speaking this stage begins with the rise of the natural sciences in the nineteenth century. Here for the first time documentation appears as a separate function. The barrier between the researcher and the material now threatens to become actual. The research atmosphere of the special library is now only to be found in the smaller departmental libraries; but the researcher must draw a part of the material from larger central libraries, where, in the main, no direct stack access is possible and where the catalogue and loan system imposes a bar between borrower and book.

(b) Supplementing the bibliographic material in this stage are lists of abstracts (of books and periodicals) in existing or specially created journals. In stage B the library catalogue is no longer the mainstay, as the emphasis is now on periodical literature.

(c) This is the position in university faculties of natural sciences today. The great difference between Stage B and Stage A lies in the existence of competing sources of knowledge: laboratory work and fieldwork. These claim the primary and greatest attention of the researcher; the literature takes only second place and is occasionally neglected. Therefore there is a tendency at this stage to delegate a part of the work to documentalists, which was not so in Stage A. Documentalists, it is felt, should not support this kind of delegation, since seldom, if ever, can they read the periodicals with the eyes of the researcher. As both Groeneveld (1947) and van der Wolk (1946) have said, the most suitable literature searcher is naturally the researcher himself. The most important function of the documentalist in this field is the help given in providing access to the literature—that is, in manipulation of the retrieval keys, which are more numerous and varied than in the humanities. Typical of the situation is, for example, the fact that the historical department of the State Agricultural University in Wageningen buys its own bibliographic aids (bibliographies of bibliographies), although most of the other natural science departments leave this to the central library. The documentalist should be superior to the researcher not only in knowledge of bibliographic aids but also in having time for documentation work and in job-satisfaction in this field (van der Wolk, 1946). Where research work serves a commercial organisation, it is specially important that the time of the documentalist (mostly of mid-grade technologist or equivalent training) be less costly than that of the research worker concerned. Here mention must be made of a literature review by Stevens (1950) which discusses the attitude of the American researcher to documentation aid (in this case from the library); the same attitude can be seen in the Netherlands. Stevens states, in fact, that the researcher wishes to see the librarian only in his 'library capacity' but not in his 'subject capacity'. The opposite point of view from the Anglo-American library and documentation world is given by Gilman (1951) and Rush (1953), who in discussing the vocational prospects drew attention particularly to the great possibilities still in store for subject librarians.

In Stage *B* documentation flourishes for the first time if not yet fully. Just as 50 years ago the librarian stalked through his rooms with his bunch of keys to the bookcases, so has the modern documentalist a paper 'bunch of keys' in the form of, for example, a card system or an index to recent accessions of the literature.

(d) Instruction in documentation of the literature is given mainly by people working in the library and documentation.

STAGE *C*—EXPERIMENT AND APPLICATION TAKE FIRST PLACE

(a) This stage begins with the application of the natural sciences to practice (agriculture and technology). Documentation here attains its full development.

(b) The barrier between research worker and material is here very marked. Research workers in technical and agricultural subjects do visit libraries but mostly they are under such pressure that they only too gladly pass on time-consuming activities such as documentation work. Even in the use of bibliographic material the barrier can be felt. Acceptance of an annotated list of the literature or a critical survey means that the original is often not consulted. The dangers of this great division between research worker and material and the resultant increased responsibility of the documentalist are clearly formulated in the report of the Netherlands Documentation Committee (Verslag (Report), 1944, p. 11): 'The preparation of a literature list on a particular subject evaluation without title based on a study of the original documents themselves can only be called poor documentation.'

(c) This is the situation today in the applied sciences (e.g. technology and agriculture). Lack of time and aversion to paperwork force people to rely almost entirely upon the help of documentation. The competing sources of knowledge are increased by one important factor: practical experiments or, in the industrial field, on a semi-technological scale. Contact with the documentation services is therefore close; in fact, there is often constant co-operation.

The growing need of the research worker for delegation of work on the literature from Stage A through Stage B to Stage C is clearly reflected in the results of the investigation carried out by Herner (1954). Of the research workers in Stage C, 29 per cent based their work on bibliographies compiled solely for their use by libraries, compared with 15 per cent of the scholars in Stages A and B. Other communications by the same author indicate that inquirers in the field of pure research occupy themselves much more fully with the literature.

It was once said quite bluntly that the 'growing need of delegation' meant nothing less than that the research workers in applied science were ever less inclined to tackle the literature 'raw'.

In regard to experience gained from practical work in Stage C, reference can be made to the publications of van der Wolk (1946) and Groeneveld (1947). In theory the importance of literature research is recognised but in practice it is the last on the list of priorities. The attitude towards documentation commonly found in research circles can be clearly seen in the following quotation (Verslag (Report), 1953): 'The working party was of the opinion that in this particular phase (namely that of orientation) care should be taken not to frighten off the research worker with the necessity for an intensive literature search.'

Allen (1969) has investigated the flow of information within R & D departments and from outside sources and has identified the existence of what he calls 'technological gatekeepers'—people who assume the role of principal communicators for teams or groups of workers.

(d) Instruction. This is entirely left to the documentalists.

STAGE D—DATA INSTEAD OF LITERATURE

Strictly speaking, this topic lies outside the terms of reference of this study, because what is here described is no longer of importance to the research worker and, secondly, it is less documentation of the literature than data documentation. However, since the expression 'documentation' is used for both literature documentation and data documentation, which often leads to confusion, it seems nevertheless necessary to say at least something briefly.

(a) Historically, documentation has always been present. Actually it is the very stuff of dissemination of information. Data documentation is as old as human society and documentation of the literature is but one of the many sources of knowledge. The report of the Netherlands Documentation Committee even traces the origin of the concept of documentation back to this. It also states that almost daily the attempt is made to use literature documentalists as disseminators of results, this occasioned by the use of the word 'documentation' for both forms.

The barrier between user and material has become insuperable, since the user is quite unwilling to consult the publication and frequenting the library is far too time-consuming.

(b) There is no need to mention bibliographic material and its compilers in this context. The consumer expects the material to be already prepackaged. The documentalist of a large company once said that it would mean dismissal if, in answer to a request from the Board for documentation, a short list of the literature were offered: the information itself was required immediately and there was no time for reading, let alone a visit to the library.

(c) Today outside the research circle this situation is found: the category of users from daily life covers a wide range from boards of directors of firms down to journalists and after-dinner speakers. They are unconcerned about being dependent upon the documentalist; they have no very high opinion of reading and study, have no time (although, sometimes, money) and see in the librarian or documentalist a general assistant.

VIEW SUMMARISING THE STAGES

In conclusion it may be said that, proceeding from Stage A through to Stage D, there is a growing emphasis on the practice; the distance between the consumer and the material widens; and interest in the literature as a source of knowledge diminishes, with faith in documentation growing ever more implicit.

This conclusion could be illustrated by analogy with, e.g., feeding in agriculture: from common pasture, A, to restricted pasture, B; then to the feeding of grain concentrates, C; and finally to the most elaborate individual feeding.

Declarations such as that of Schischhoff (1952) on the future of documentation in such fields as philosophy would find, in the author's view, but little support. Present conditions appear to be static and there is no sign for the moment that, in the interplay between research worker, documentation and library, any important shift can be expected. It can be hoped that the confidence of the scholars described in Stage A and the 'know-how' of the documentalist in Stage B may increase. It has been seen that the self-reliance and confidence shown by the A group does not foster confidence in documentation; and this confidence is lacking in the B group also, although it is growing, since documentation is as yet too immature to be able to assume a clearly fixed position. Confidence can be won by documentalists by devotion to their assignments and awareness of possibilities of training and refresher courses. In Stage C documentation is much more accepted, so that here little further development can be expected; users can easily calculate that delegation of literature searching to an experienced documentalist is always more advantageous.

The social sciences do not fit easily into this division into stages. History and sociology come under Stage A; environmental studies (with statistical surveys) perhaps more under Stage B. Economics can be theoretical (Stage A) but can also be regarded as practical.

In this connection, mention should be made of the book by Rothstein (1955). This history of reference services in American libraries contains a certain amount on the history of delegation of literature surveys in research libraries. Among other things, a detailed description is given of the appointment of subject specialists as helpers in literature searching in the Library of Congress before World War II. The then Librarian, Putnam, introduced three levels in all (chairs, consultants and honorary consultants) exclusively charged with this work. Rothstein cites much literature on the subject and also reports opinions which are not in agreement with delegation of literature searches.

References

Allen, Th. J. (1969). 'Information Needs and Uses.' *A. R. Inf. Sci. Technol.* **4**, 4

Alphen, G. van (1953). 'Die Universitere onderwijs van Nederlandse Kultuurgeschiedenis in Suid Afrika.' *Standpunt No. 5*

Eppelsheimer, H. W. (1961). 'Die Dokumentation in den Geisteswissenschaften.' *Nachr. Dokum.* **2**, No. 3, 87

Evans, M. G. (1956). 'Duplication in Experimental and Theoretical Disciplines and its Implications for Modern Documentation.' *Am. Docum.* **7**, No. 1, 46

Fairthorne, R. (1969). 'Content Analysis, Specification and Control.' *A. R. Inf. Sci. Technol.* **4**, 73

Gilman, E. R. (1951). 'Science Librarians Wanted.' *Libr. J.* **76**, 1854

Groeneveld, C. (1947). 'Research in Bedrijfsbibliotheek.' *Bibliotheekleven* **32**, 153, 173

Herner, S. (1954). 'Information Gathering Habits of Workers in Pure and Applied Science.' *Ind. Engng Chem.* **46**, 229

Rothstein, S. (1955). *The Development of Reference Services through Academic Traditions, Public Library Practice and Special Librarianship.* Chicago; A.L.A.

Rush, N. O. (1953). 'Service to Readers of University Libraries.' *Libr. Ass. Rec.* **55**, No. 10, 213

Schischhoff, G. (1952). 'Über die Möglichkeit der Dokumentation auf dem Gebiet der Philosophie.' *Nachr. Dokum.* **3**, No. 1, 20

Stevens, R. E. (1950). *A Summary of the Literature on the Use made by the Research Worker of the University Library Catalogue.* (Occasional Papers No. 13.) Univ. of Illinois Libr. School

Verslag (1944). 'Verslag van de Documentatie commissie van de Nederlandse Vereniging van Bibliothecarissen.' *Bibliotheekleven* **29**, No. 19, 21

Verslag (1953). 'Beknopt verslag van de werkgroep I voor het onderwerp Organisatie problemen, welke verband houden met research werk.' *De Ingenieur* **65**, A443–6

Wolk, L. J. van der (1946). 'Literatuuronderzoek in het bedrijf.' *Bibliotheekleven* **31**, 159–67

NINE

THE USE OF BIBLIOGRAPHIC AIDS

INTRODUCTION

It has been shown that literature research, originally an integral part of scholarly research, for many reasons and in differing situations can be, and will be, delegated; librarianship and documentation are thus faced with a new task, which has been designated passive documentation.

Where the literature search is undertaken by the scholar himself as a part of his academic research, it naturally follows an individual search pattern, the scholar's individual 'search habits.' These search habits are very important, for those who are entrusted with a literature search, as work methods to be recommended; at the same time, investigation into search habits of scholars is a sound foundation for appraisal of the practical uses of the results of active documentation. In other words, the main benefit derived from an assessment of search habits is the acquisition of information, from the point of view of librarianship and documentation, about the use of expensive bibliographic aids.

Before going on to the assessment of search habits it is worth paying attention to the relative importance of literature activity in research work as a whole. It is important to appreciate that a research project can not be completed through literature research alone.

ASSESSMENT OF INFORMATION GATHERING HABITS

Investigation of search habits is important:

1. As a foundation for 'active' documentation.
2. As a guideline for literature search delegated to librarianship and documentation.

Bibliographies on the investigation of search habits are given in Davis and Bailey (1964), Slater (1968) and Wood (1971). Moreover, the subject is reviewed annually in the volumes of the *Annual Review of Information Science and Technology*. Of interest is the growing importance of this area of research in the field of Sociology of Science and in the behavioural sciences.

In many investigations into search habits the chain of related problems is extended to include communication habits in all branches of knowledge. The result is often that the importance of literature research within the totality of communication rates lower than is willingly admitted by librarians and documentalists. Oral communications, browsing, own research, and so on, are named as attractive alternatives to literature research.

An important review article is that of Tornüdd (1959), who examined 72 previous investigations, carried out as pilot work for an investigation on the use of scientific literature and reference services by Scandinavian scientists and engineers.

An indication of the increased interest in this field is the fact that Paisley (1968), reviewing the pertinent literature in the year 1967 only, collected *also* 70 investigations (mainly American). Research is stimulated by the supposed co-ordination of a scientist's productivity with his utilisation of communication channels (Parker, Lingwood and Paisley, 1968).

The main problems encountered in this research are as follows (the first four are external and the last two are internal):

1. Diversity of motivation for literature searching (orientation, retrospective searching, current awareness).
2. Availability of bibliographic aids in the immediate environment.
3. Discipline of the searcher (humanities, natural sciences, techniques) (Herner, 1954).
4. Environmental influences (country, cultural patterns, university, government, industry).
5. The difficulty of describing the searching behaviour.
6. The measurement of the relevancy of the documents produced (see Cuadra and Katter, 1967).

The methods used in investigations of search habits were as follows.

Direct methods

1. Questionnaire or interview method.

2. Diary books analysis.
3. Controlled experiments.

Indirect methods

1. Collection of evidence and observation of research workers on the job.
2. Loan statistics.
3. Statistical treatment of references.

As good examples of statistical treatment of references in the literature, Stevens (1956) gives the investigations of Fussler (1949) and McAnally (1951). This method is more 'objective', since not so many subjective elements can infiltrate as in the questionnaire. Against this it has narrower limits. What can be measured? The following can be established:

(a) In which years most of the quoted articles appeared.
(b) Number of titles deviating from main subject.
(c) Frequency distribution of languages.
(d) Frequency of various forms of the literature.

This method does not, however, say *in what way* the researcher came to the literature quoted. According to Stevens, account must

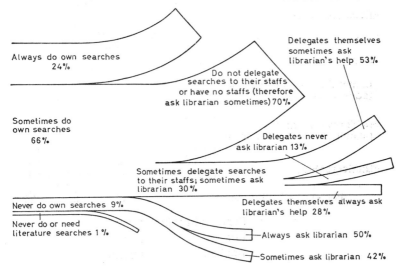

Figure 13. Literature search habits. (Hogg and Smith, 1958, by courtesy of Proc. Int. Conf. Sci. Inf.)

be taken of the criticism of this method by, e.g., Brodman (quoted by Stevens). Brodman found a high correlation between the frequency counts of references to certain periodicals and counts were a measure of the importance attributed to the periodicals by subject specialists. She reaches the conclusion that more comprehensive and thorough research in the field of librarianship and documentation is a pressing necessity.

Results

Owing to the considerable number of investigations, we can mention only a few of the results. This selection is purely random and does not imply any criticism of the investigations that are not mentioned.

1. Hogg and Smith (1958) produced a very instructive diagram (*Figure 13*).
2. Herner's diagram (*Figure 14*) is interesting also, because of the differentiation by environment.
3. Survey (1965) contains some remarkable conclusions:

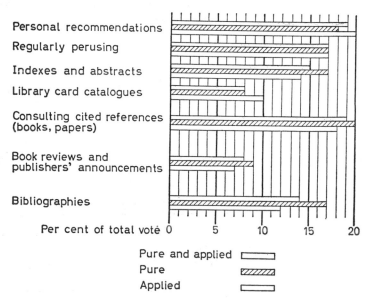

Figure 14. Relative use made of indirect sources of information. (After Herner, 1954)

(a) The respondents have *not* the usual preference for verbal communication.
(b) They request more review articles.
(c) They urge the introduction more widely of a system of indicating within a paper what is new, what is introductory and what is experimental procedure (see also p. 97).
(d) They suggest more and better information on the use of their libraries.

4. Martyn (1964) arrives at the following results by means of a questionnaire:

80 per cent follow up references cited in relevant papers,
77 per cent 'keep up' by reading current publications,
66 per cent gain references from conversations with other workers in the field,
58 per cent use indexes in journals, including abstracts journals,
50 per cent use standard textbooks and monographs,
47 per cent use a personal index or record,
31 per cent gain references from correspondence,
26 per cent try to obtain unpublished material from other workers,
16 per cent use a subject bibliography,
14 per cent consult reports of internal circulation,
10 per cent use a library card index,
 8 per cent ask a librarian or information officer.

FOUR DIFFERENT APPROACHES TO INFORMATION GATHERING

From the results of research on information gathering habits and from oral reports by librarians and documentalists on European practices, four approaches can be distinguished:

1. Preference for oral communication.
2. Preference for following only a few 'star' workers, 'star' institutes and 'star' periodicals.
3. Preference for the use of bibliographic coupling, i.e. documents related by common citations.
4. Preference for the practising of a systematic literature search with the help of bibliographic aids.

These approaches are not mutually exclusive but rather supplementary. They have their optimal applicability for:

(a) Quick information on parts of specific references.

(b) Orientation in an area new to the inquirer.
(c) Research on the relations between individual researchers and their work. This is important in the field of patents, in research on the history of an idea, etc.
(d) Thorough retrospective literature searching.

Oral communication is considered more important in the U.S.A. than in Europe. Menzel (1968) and de Solla Price and Beaver (1966) indicated that 80 per cent of all communication between scientists is handled through oral or unofficial written channels. Also, Garvey and Griffiths (1968) found that a considerable amount of information reaches scientists through so-called informal channels. The very interesting account by Watson (1969) of his information gathering methods during his work on the structure of DNA illustrates vividly the importance of communication by speech and letters rather than through formal documentary channels. Naturally, oral communication has the advantage of promptness, spontaneity, specificity, personal interaction, etc. However, the formulations tend to be less concise and correct and the information has not been scrutinised by a controlling board of editors. Thus the scientific community maintains an invisible network, for which de Solla Price coined the phrase 'invisible colleges'. Formalisation of these invisible colleges into so-called Information Exchange Groups, e.g. within the National Institutes of Health, enjoyed a brief vogue, but was abandoned after a few years because of the threat of boycott by commercial publishers, among other reasons.

Orientation to 'star' workers, 'star' institutes and 'star' periodicals is a method of avoiding the drudgery of thorough systematic literature searching. The evaluation of individuals or institutes is a subjective matter, but the 'star' status of periodicals and serials can under certain circumstances be quantified by such things as the number of times articles from the periodical in question are cited elsewhere. These statistics may be obtained from a study of *Science Citation Indexes* (see p. 84). This so-called phenomenon of scattering and the concentration of the most important material in a relatively low number of periodicals has already been touched upon, and the cumulative curves of Taube have been discussed (see Chapter 4, p. 34). In a non-cumulative form these curves are similar to the hyperbolic curve of Zipf–Estoup ($xy=k$). The applicability of this curve of Zipf–Estoup in the situations of interest to us is discussed by Fairthorne (1969), who gives a review of the literature in this field.

Concentration has certainly taken place and to some extent justifies the use of the 'star' method; as already stated, however, it is really valuable only in a situation requiring an original orientation.

Approach (3) will be discussed later in this chapter, and approach (4) is discussed in the next section.

SYSTEMATIC LITERATURE SEARCHING

In this section we discuss systematic working-up of the literature: that is, following a fixed path in the use of bibliographic aids, as described below.

Figure 15 collates the bibliographic material mentioned in *Figure 4* (for articles in periodicals) and *Figure 11* (for separate publications).

First of all, definite characteristics of the bibliographic material are to be examined. These characteristics vary in the individual sectors. In the case of material in the MU sector, the starting point is the producer. Most of the material is divided by subject heading, little by classification. Strict division is observed according to the form of the material; that is, by periodical articles on the one hand and separate publications (books) on the other. Material in the LI sector is compiled from the point of view of the collection. Many other classifications exist in addition to the subject heading division. The influence of accidental form is great and the working-up of book and that of periodical material are normally entirely distinct operations.

Material in the DO sector is dealt with from the subject content viewpoint, according to classification or by subject heading, and the form of publication plays no part at all. Material in the RE sector, too, is compiled only for subject content, form being unimportant; it is very seldom arranged by subject heading, mostly by special classifications.

The properties of this bibliographic material can now be grouped as in Table 5.

If the arrow in *Figure 15* is followed, it will be seen to reach

Table 5

Sector	Starting point in producing the material	Division	Influence of form of publication
MU	Producer	Subject heading	Yes
LI	Collection	Subject heading and class	Yes
DO	Subject matter	Subject heading and class	Sometimes
RE	Subject matter	Classified	None

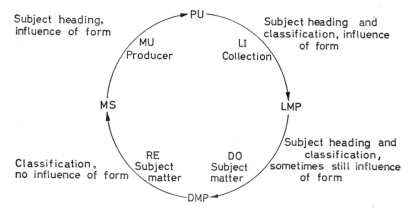

Figure 15. Properties of bibliographic material

from the MU sector, through sectors LI and DO to the RE sector.
This sector provides the ideal form of bibliographic material for
literature search: it is always in classified order and the accidental
form (book or the periodical article) no longer counts.

In the case of systematic working-up of the literature, it is best to
begin with material in the RE sector and to follow its development
in the direction of the arrow in *Figure 16*; that is, in reverse. For
such a literature search the following order is achieved:

1. Key publications, which provide a sound foundation.

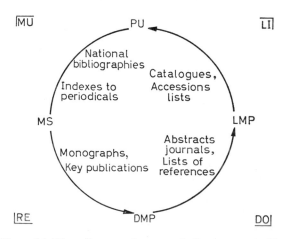

Figure 16. Flow diagram of systematic literature searching

2. Abstracts journals, which cover that part of the literature for which no key publications exist.
3. Library catalogues, which help, above all, to locate the material and possibly add to the references discovered in (1) and (2).
4. In exceptional cases national bibliographies or indexes to specialised periodicals can help. National bibliographies may enable one to obtain the first key publications (in this case, books only).

USE OF THE CITATION RELATIONS BETWEEN DOCUMENTS

There are two kinds of relations:

1. Relations by citation going from the *citing* publication to the *cited* publication (so-called snowball system; see below).
2. Relations by citation going from the *cited* publication to the *citing* publication (so-called Citation Indexes; see p. 84).

The Snowball System

In practice, use is often, indeed mostly, made of the so-called snowball system: one starts by following up the references quoted in a recent article and then the references found therein and so on. This has the advantage that the literature is quickly covered. About the disadvantages more will be said later.

The popularity of this system with research workers can be seen in, for example, the investigations of Herner (1954), Bernal (1948) and Urquhart (1948). A questionnaire has shown, as already mentioned above, that references were more valued than abstracts journals.

Let us take a hypothetical example to illustrate this system: 'Jansen' (1950) is credited with having published in a Netherlands periodical in the Dutch language. In *Figure 17* the Y-axis gives the years; the X-axis, the authors by language and alphabetically arranged within the language (American and English authors under the heading English). For the sake of simplicity there is a limit of three languages. As a possible variation, instead of the authors, periodicals could be set out on the X-axis. 'Jansen' (1950) quotes three articles, namely two Dutch, 'Gerritsen' (1948) and 'Pietersen' (1946), and an English article, 'Wood' (1948). The references are shown joined up (reference links). In the follow-up it is found that

'Gerritsen' (1948) does not give any other reference and that 'Pietersen' quotes an article of his own from 1948 and adds that a survey of the literature can there be found. He also names a dated German review article by 'Müller' (1933). 'Wood', for his part, quotes another two authors, 'Smith' (1941) and 'Jones' (1938); of the two, 'Jones' had given an exhaustive literature survey.

It must be strongly emphasised that the illustration is *very much* simplified for reasons of clarity and that, in practice, the network of reference links would be much larger and more complicated. In the case of key articles, the coverage of a specific language area is indicated by a stroke–dot line. Complete application of this 'snowball' system, therefore, leads to the discovery of a network of reference links connecting the publications together.

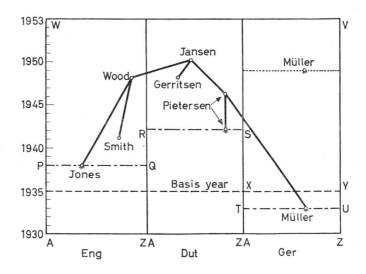

Figure 17. The 'snowball' system

By use of the snowball system specific key publications are run to earth in the majority of cases (monographs, surveys, reviews, etc.) which spare the researcher the trouble of further search in that particular language area. In theory, key publications can cover more than one language area—indeed, all language areas. Practice shows, however, that accuracy and comprehensiveness may only be expected in the author's own language. Some relevant statistics are given in an article by Blanchard (1950), who quotes percentages of references from the English language area in an American publication from 64.5 per cent up to 92.6 per cent. Another important paper is that of Visscher (1954), who cites the following causes of 'provincialism' in references:

(a) Language boundaries.
(b) Inadequate funds for foreign acquisitions.
(c) Cultural idiosyncracy.
(d) Self-advancement motives.
(e) Proximity and accessibility.

Visscher calculated the percentages of provenance of articles quoted in *Physiological Reviews* and *Biological Abstracts*.

Another example of 'provincialism' is afforded by Mitchell (1956), who quotes the following figures in a letter to the editor of *Science*. In 100 books by American geologists he found 75 with less than 10 per cent of non-American references, 18 with 11–20 per cent, 5 with 21–30 per cent, 2 with 31–40 per cent, 0 with 41–50 per cent and 0 with more than 50 per cent. The same counts were applied to foreign citations in a further 100 English books and 50 German, French, Dutch and Swiss books: the astonishing results presented in Table 6 were arrived at.

Table 6 (after Mitchell, 1956)

	<10%	*11–20%*	*21–30%*	*31–40%*	*41–50%*	*>50%*
(100) U.S.A.	75	18	5	2	0	0
(100) England	11	36	33	9	5	6
(50) Germany	8	54	18	8	4	8
(50) France	19	24	30	18	4	0
(50) Switzerland	14	12	18	22	14	20
(50) Netherlands	4	8	16	8	16	48

It can be seen that the smaller the country, the less provincial the authors can afford to be. The influence of the discipline on provincialism is shown by Stevens (1953). (See Table 7.)

Returning to the hypothetical article by 'Jansen' (1950), three key publications are found—'Jones' (1938), 'Pietersen' (1942) and 'Müller' (1933)—which together form the so-called frontline of key publications (line PQRSTU). From this base further probes can be made into the as yet only partially explored territory PQRSTUVW.

Let it be supposed that publications prior to 1935 were only of limited value and that for this reason 1935 would be taken as the basis year (*Figure 17*). This would mean that 'Jones' would save any search in the English language area from 1935 to 1938; 'Pietersen', any search for the Netherlands from 1935 to 1942. 'Müller' (1933) is obsolete. Cover would thus be provided in the English language up to 1938 and in Dutch until 1942; but there would be nothing in the German language, entailing a search in the German language area from 1935 to date.

Table 7 (after Stevens, 1953)

Publication	Discipline	Percentage of references			
		English	German	French	Others
Allen (1930)	Mathematics	66.6	20.0	16.8	6.5
Sheppard (1935)	Chemistry	46.1	38.8	7.8	7.3
Cross and Woodford (1931)	Geology	82.0	11.6	2.0	4.4
Fussler (1949)	Chemistry	64.5	25.0	3.0	7.5
Hooker (1935)	Physics	67.4	26.7	3.0	2.9
Fussler (1949)	Physics	66.6	22.1	2.9	8.4
Hooker (1935)	Radiotechnology	72.4	17.1	6.0	4.5

In other words, after the front of the key publications has been fixed, all that remains is systematic working-up of the literature in the area PQRSXYVW in *Figure 17*. In this area some publications have already been found. The dangers of the comparatively arbitrarily chosen starting point can thus be seen. At the beginning one might have only been cognizant of, for example, 'Pietersen' (1946) or an article from 1953, now missing. It must always be borne in mind that other, not quoted but nevertheless extant, works in the area PQRSXYVW might be quite unknown; everything in this area above the 1950 line cannot possibly be covered by 'Jansen' (1950) and everything under the line has only become visible to a small degree. It is indeed possible that key publications may be missed in this way; for example, 'Müller' could well have published in 1949 a fully revised edition of his previous survey. The snowball system in itself must always lead to unsatisfactory results unless a very recent review article is available in a specific language area. Yet there are authors, such as Baer (1959), who recommend the snowball system and set no value on coverage in abstracts journals.

A further major disadvantage of the snowball system is that one rapidly arrives at the older literature. According to Donker Duyvis (1946), there would be fully 70 per cent of the required articles, in a documentation service, not older than five and 90 per cent not older than ten years.

Stevens (1953) has shown that these figures are not the same for all branches of knowledge. His findings are set out in Table 8.

Table 8 (after Stevens, 1953)

Publication	Discipline	Percentage of articles not older (in years) than				
		2	5	10	15	20
Sherwood (1932)	Medicine		55	75		
Hunt (1937)	Chemistry		52.1	73.6	84.9	89.4
Jenkins (1931)	Medicine			82		93
Fussler (1949)	Chemistry	30.2	51.3	71.3	78.7	
Fussler (1949)	Physics	40.4	69.4	88.2	93.9	
McAnally (1951)	History of U.S.A.		10.4			33.8

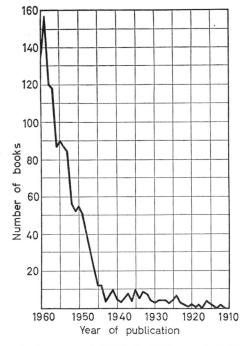

Figure 18. Books lent from Yale Medical Library plotted by publication date. (After Kilgour, 1961, by courtesy of American Documentation)

Kilgour (1961) shows clearly in a diagram (*Figure 18*) the interest of medical workers in the latest literature. In this case, too, ageing of the literature is seen to lead to rapid loss of value. Burton and Keller (1960) also gave a conspectus of the rapid ageing ('half-life') of publications in various fields.

Bourne (1963) states that—on the basis of 28 investigations about ageing publications—he can offer no generalisations on the so-called 'half-life'. Mikhailov, Chernyi and Gilyaverskii (1965) mention the quick ageing of publications in the fields of physics, chemistry, geology and physiology, compared with publications in botany, zoology and entomology. There is a general tendency towards 'recent cites recent' and older publications (with the exception of reviews) are soon forgotten (de Solla Price, 1965). Parker, Paisley and Garrett (1967) suggest that disciplines can be compared through citation analysis, and that emerging disciplines show fewer and more closely related citations.

Objections to the snowball system can be summarised as follows :

(a) Dependence upon the care shown by authors of starting-point articles in the matter of literature gathering.
(b) Rapid retrieval of older (obsolete) literature.
(c) Account to be taken of author's provincialism.

The snowball system by itself can never afford complete coverage; it must be supplemented by systematic literature searching. On the other hand, systematic literature searching should be supplemented by the snowball system. The grounds for this will now be discussed.

Despite its shortcomings, the snowball system is essential as an adjunct to systematic literature searching. The authors' reference lists contain the results of their personal contacts with colleagues working in the same field and the exchange of offprints is a very common method of contact between research workers. All the same, Urquhart (1952) says that researchers always obtain more information through libraries than through the exchange of offprints. Thus there is a second circle which completely by-passes the previously mentioned circle (p. 83): MU—offprint—RE (*Figure 19*).

It may be asked how researchers obtain the addresses of colleagues. The answer is: through personal contact at conferences or study tours, from reports of institutes, from the professional press and so on—and, of course, researchers will be well informed about activities and personalities in their specialist fields. In this way the documentalist, too, should be able to gather material through his knowledge of the sources in the realm of research, the 'geography of research'. This knowledge can be based on: address books of research centres, lists of conference participants, research programmes, periodical indexes and, particularly, special publication series of institutes. This geography of research is the easier to explore the more costly, and therefore the rarer, the 'hardware' required for that particular field of research. Addresses where nuclear energy research is carried out, for example, are generally well known. Special laboratories, too, are mostly so expensive that each country can afford but a few.

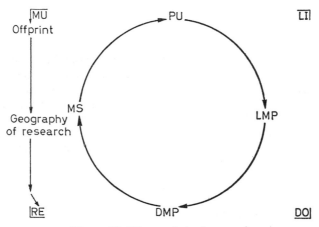

Figure 19. 'Geography' of research

The position is entirely different in the field of the humanities. An individual can mount a project unaided by outside finance, if no long journeys need to be undertaken. Any student of the history of art living in the Netherlands, if he has a mind to write a biography of Frans Hals, would need but little money, given the scholarly education and training: the books can mostly be borrowed free from libraries and there is no need for trial grounds and laboratories. In the world of learning he can remain absolutely unknown until the appearance of his book. This means, as far as documentation is concerned, that the more costly the experimental infrastructure, the better known the research addresses and the easier the documentation. Systematic literature searching can therefore be supplemented by coverage of the geography of research.

Velander and Hellström (1955) describe developments in this field in Sweden: research workers in each working syndicate have a contact man who keeps in touch with a central contact point and disseminates and receives, respectively, information in his field (on demand or not). According to the two authors, people prove to be much less difficult than official agencies in passing on information. These reference services are a consequence of the lack of activity in documentation centres: it is just a matter of 'do it yourself'. Finally, Velander and Hellström give a list of addresses of research workers in different countries in various fields. That is just what is meant by 'geography in research'; in Sweden the research workers do it for themselves. Holt (1960) provides a good example of the interest aroused in a documentation centre by the 'geography of research'.

Conclusion

The researcher is dependent on many things in the literature search: on the acquisition and cataloguing policy of librarians; on the coverage policy of editors of abstracts journals; on the care exercised by his predecessors, especially the compilers of comprehensive review articles and monographs; and finally on all writers quoted by him, in so far as he re-uses their lists of references employing the snowball system. The conclusion can be drawn that comprehensive literature searching should cover:

(a) Key publications.
(b) Abstracts journals.
(c) Library catalogues.
(d) National bibliographies and indexes, optional.
(e) Geography of research; always in conjunction with the snowball system.

All types of possibilities could be mentioned on a search form to ensure that the search proceeds systematically and that a reliable comprehensive result is achieved. Such search forms can be prepared according to circumstances in the institutes concerned.

Citation Indexes

A citation index consists of two different parts:

1. An index of all articles published in a selected group of periodicals in a given year (so-called sources index).
2. An index, arranged by author, of all articles cited in the articles of group (1) (so-called citation index).

The use of a citation index is: locating a known author in (2) and searching the sources, quoting his articles, in (1). If no author is known, a subject index is available leading to author names.

Citation indexes have entered the field of science with the publication of the *Science Citation Index* by Garfield (1955, 1968). The idea is not new: *Shepherd's Index* to legal cases started in 1873. The revival has been stimulated by the availability of computers as aids in compiling and arranging the indexes in question.

One of the basic ideas is that only 2200 of the most important periodicals (as covered by *Science Citation Index*) out of a population of 35 000 ensure a satisfactory coverage due to the Zipf distribution already discussed (p. 74).

The principal use of citation indexes is in:

(a) Searching for the history of an idea (patent).
(b) Searching for the use and expansion of a certain method after its first publication.
(c) Searching in multidisciplinary fields.

If citation indexes are used alternately with the snowball system, the process is called 'cycling'.

Two main advantages are the absence of terminological problems and the rapidity of searching compared with the use of abstracts journals.

The main disadvantage is that citation relations between documents are more often relations between microthemes (see Chapter 10) of these documents than between the documents as a whole. This is not a difficulty in the snowball system, which leads from microtheme to microtheme, but it is in citation indexes, which lead from a microtheme to a complete document only. The second step, from this document (found with the first step) to another document, entails a rather disturbing amount of noise.

Useful comparisons between results obtained with abstracts journals on the one hand and citation indexes on the other have been made by Spencer (1967), Martyn (1965) and Huang (1968). Generalising these comparisons, one can say that the two methods supplement each other and should be used simultaneously. Holt (1960) can be mentioned as a practical pioneer of the use of citation indexes.

The publication of science citation indexes has led to some additional research:

1. *Statistics.* Citation indexes enabled research workers to count citation frequency of periodicals (Zipf distribution: titles of importance for both editors of abstract journals and library acquisition), of authors and of articles. The relation of the two last frequencies to 'importance' has, as a matter of fact, been doubted.

2. *Communication patterns.* More rewarding has been the use of citation indexes for research in the field of patterns of communication between research workers, especially between those working on the front of research (de Solla Price, 1965; Parker, Paisley and Garrett, 1967). They have also been used for measuring interdisciplinary relations, e.g. between biology and medicine.

3. *Bibliographic Coupling.* Related research is based on bibliographic coupling, a term originated by Kessler (1963, 1965). The idea is that documents containing a certain number of the

same literature references are probably more or less of the same field: this could lead to the isolation of clusters of related documents in a population of documents.

References

Baer, K. A. (1959). 'Bibliographical Methods in Biological Sciences.' *Spec. Libr.* **45**, No. 2, 79

Bernal, J. D. (1948). 'Preliminary Analysis of Pilot Questionnaire on the Use of Scientific Literature.' *Report and Papers Roy. Soc. Scient. Inf. Conf. London.* (Paper No. 46), pp. 589–637

Blanchard, J. R. (1950). 'Agricultural Research and the Exchange Problem.' *Coll. Res. Libr.* **11**, 40, 53

Bourne, C. P. (1963). *Methods of Information Handling.* New York; Wiley

Burton, R. E. and Keller, R. W. (1960). 'The "Half-life" of Some Scientific and Technical Literature.' *Am. Docum.* **11**, No. 1, 18

Cuadra, C. A. and Katter, R. V. (1967). 'Opening the Black Box of "Relevance".' *J. Docum.* **23**, No. 4, 291

Davis, R. A. and Bailey, C. A. (1964). *Bibliography of Use Studies.* Philadelphia; Drexel Inst. of Technology: Graduate School of Library Science

Donker Duyvis, F. (1946). 'List of Periodicals for Users Specialized in Science and Technology.' *F.I.D. Comm.* **13**, No. 3, C22–5

Fairthorne, A. (1969). 'Empirical Hyperbolic Distributions (Bradford, Zipf, Mandelbrot) for Bibliometric Description and Prediction.' *J. Docum.* **25**, No. 4, 319

Fussler, H. H. (1949). 'Characteristics of the Research Literature Used by Chemists and Physicists in USA.' *Libr. Q.* **19**, 19

Garfield, E. (1955). 'Citation Indexes for Science.' *Science, N.Y.* **122**, 108

Garfield, E. (1968). 'World Brain or Memex.' In: Montgomery, E. B. (ed.). *The Foundations of Access to Knowledge.* pp. 169–196. Syracuse; Syracuse Univ. Press

Garvey, W. D. and Griffiths, B. C. (1968). 'Informal Channels of Communication in the Behavioral Sciences.' In: Montgomery, E. B. (ed.). *The Foundations of Access to Knowledge.* pp. 129–146. Syracuse; Syracuse Univ. Press.

Herner, S. (1954). 'Information Gathering Habits of Workers in Pure and Applied Science.' *Ind Engng Chem.* **46**, 229

Hogg, I. M. and Roland Smith, J. (1958). 'Information and Literature Use in a Research and Development Organization.' In: *Proc. Int. Conf. Sci. Inf., Washington*

Holt, S. J. (1960). 'The Intelligence Service for Aquatic Sciences and Fisheries Provided by the Food & Agricultural Organization of the U.N.' *Revue Docum.* **27**, No. 3, 108

Huang, Th. S. (1968). 'Efficacy of Citation Indexing in Reference Retrieval.' *Libr. Res. Techn. Survey* **12**, No. 4, 415

Kessler, M. M. (1963). 'Bibliographic Coupling between Scientific Papers.' *Am. Docum.* **14**, No. 1, 10

Kessler, M. M. (1965). 'Comparison of the Results of Bibliographic Coupling and Analytical Subject Indexing.' *Am. Docum.* **16**, No. 3, 223

Kilgour, F. G. (1961). 'Recorded Use of Books in the Yale Medical Library.' *Am. Docum.* **12**, No. 4, 266

McAnally, A. M. (1951). *Characteristics of Material used in Research in U.S. History* (unpublished *Ph.D. Thesis* Graduate Library School, Univ. of Chicago), quoted by Stevens

Martyn, J. (1964). *Report of an Investigation on Literature Searching by Research Scientists.* London; ASLIB Research Dept.

Martyn, J. (1965). 'An Examination of Citation Indexes.' *ASLIB Proc.* **17**, 6 (June)

Menzel, H. (1968). 'Informal Communication in Science.' In: Montgomery, E. B. (ed.). *The Foundations of Access to Knowledge.* Syracuse; Syracuse Univ. Press

Mikhailov, A I., Chernyi, A. N. and Gilyaverskii, R. S. (1965). *Basis of Scientific Information.* Moscow

Mitchell, R. E. (1956). 'Lack of Recognition of Foreign Works.' *Science, N.Y.* **123**, 990

Paisley, W. J. (1968). 'Information Needs and Uses.' *A. Rev. Inf. Sci. Technol.* **3**, 1

Parker, E. B., Lingwood, D. A. and Paisley, W. J. (1968). *Communication and Research Productivity in an Interdisciplinary Behavioral Science Research Area.* Palo Alto; Stanford Univ. Inst. Communication Research

Parker, E. B., Paisley, W. J. and Garrett, R. (1967). *Bibliographic Citations, an Unobtrusive Measure of Scientific Communication.* Palo Alto; Stanford Univ. Inst. Communication Research

Slater, M. (1968). 'Meeting the User's Needs within the Library.' In: Burkett, J. (ed.). *Trends in Special Librarianship.* pp. 99–136. London; Bingley

Solla Price, D. J. de (1965). 'Networks of Scientific Papers.' *Science, N.Y.* **149**, 510

Solla Price, D. J. de and Beaver, D. (1966). 'Collaboration in an Invisible College. *Am. Psychologist* **21**, 1011

Spencer, C. C. (1967). 'Subject Searching with Science Citation Index, Preparation of a Drug Bibliography using Chemical Abstracts, Index Medicus and Science Citation Index 1961 and 1964.' *Am. Docum.* **18**, 87

Stevens, R. E. (1953). *Characteristics of Subject Literature* (ACRL Monogr. No. 6)

Stevens, R. E. (1956). 'The Study of Research Use of Libraries.' *Libr. Q.* **26**, 41

Survey (1965). 'Survey of Information Needs of Physicists and Chemists.' *J. Docum.* **21**, No. 2, 83

Tornüdd, E. (1959). 'Study on the Use of Scientific Literature and Reference Services by Scandinavian Scientists and Engineers Engaged in Research and Development.' In: *Proc. Int. Conf. Scient. Inf., Washington.* Vol. 1, pp. 19–76

Urquhart, D J. (1948). 'The Organization of the Distribution of Scientific and Technical Information.' *Report and Papers Roy. Soc. Scient. Inf. Conf. London.* pp. 75–89

Urquhart, D. J. (1952). 'A Review of the Results of the Royal Society Scientific Information Conference 1948.' *ASLIB Proc.* **4**, No. 4, 233

Velander, E. and Hellström, K. (1955). 'Exchange of Documentary Information.' In: *Proc. Congr. Int. Bibl. Centr. Docum., Brussels, 1955.* Vol. 1, pp. 191–199

Visscher, M. B. (1954). 'Interdependence of Knowledge and Information in the World Today.' *Libr. Q.* **24**, No. 2, 81

Watson, J. D. (1969). *The Double Helix.* New York

Wood, D. N. (1971). 'User Studies.' *ASLIB Proc.* **23**, 11 (Jan.)

TEN

RETRIEVAL SYSTEMS

INTRODUCTION

The chapters on active documentation dealt with the bibliographic tools; in Chapter 9, on passive documentation, the use of these aids in literature searches was described. Consideration can now be given to the inner structure of these bibliographic aids, namely the systems which are employed in order to provide, *within* the bibliographic material, access to the literature, with regard to its *content*. The means used to this end are referred to below as retrieval systems.'*

Bibliographic material can consist of a compilatory part and a retrieval system, e.g. in an abstracts journal: abstracts and index. But the compilatory parts can have a built-in retrieval system, where the bibliographic material is identical with the retrieval system (e.g. indexes such as the *Biological and Agricultural Index* or in the so-called 'collectanea', where text and retrieval system are together; cf. Hines, 1961). There are also retrieval systems to material in a *single* volume (index).

The word 'retrieval' invariably evokes thoughts of 'machines'; 'retrieval system', however, is never synonymous with 'machine use' (see Chapter 18, pp. 162–163).

DEFINITIONS OF TERMS

A retrieval system consists of *elements* which are composed as follows: an information carrier (I), which carries a link between a descriptor (D) and the address (A) of a document. The simplest example is a reference in an index of a book: Tomatoes (D), page

* This expression may be misunderstood: a retrieval system may be considered as a system including document, user, enquiries and the retrieval devices. However, the phrase 'retrieval system' is generally used as a synonym for 'retrieval device', and we follow this practice in our context.

99 (*A*), where the paper is (*I*) and where the selected 'document' is a paragraph dealing with tomatoes in the book itself.

Information carriers (*I*) can be cards, tapes, film, dyeline paper or elements of a calculating machine, etc. The information carrier determines the outer form of the retrieval systems.

By *descriptor* (*D*) (cf. p. 40) is meant a word, a concept, a notation, by means of which a publication can be retrieved. In theory, all words of all languages could be descriptors. In practice, various constraints are often imposed which lead to a certain degree of standardisation. The best-known and also the most radical is the compilation of a list of chosen (permitted) descriptors, a so-called thesaurus. (Such a list often includes also non-permitted descriptors with a reference to permitted descriptors; see Chapter 12.)

The *document address* (*A*) can be *primary*. In this case the exact location of the document within a collection of material is given, as is usual in catalogues. The address is said to be bibliographic or *secondary* if an article, as is usual in a documentation system, is referred to by the title of a periodical alone. This means that the actual location of the periodical must be determined in a further system (catalogue). A *tertiary* document address would therefore be, for example, a page number in the index to an abstracts journal.

If the descriptors of a document were, for example, Tomatoes—Diseases—Netherlands, these words (*D*) together with a document address (*A*) entered on a catalogue card (*I*) would constitute what has been referred to above as an *element*. A number of such elements together compose a retrieval system. A complete catalogue card carries, for example, subject heading (descriptor), shelf location (address) and title. The title in this context is ambiguous: on the one hand, it is an extension of the information contained in the descriptor and, on the other, it contains certain address elements, especially in a documentation system.

THE LINK-UP

The link-up of documents and queries, on the one hand, with the retrieval system, on the other, can be presented in a diagram (see *Figure 20*). In this diagram three activities are mentioned (with correspondence in the query sector):

1. (See Chapter 11.) Condensation of the document without the help of a thesaurus. This may be done by three methods: (a) pure derivation from the text with no other words added—so-called methods with derivation; (b) free indexing (expanded derivation); (c) abstracting (also expanded derivation).

2. (See Chapter 12.) Reformulation of the condensate with the
help of a thesaurus—so-called methods with assignment.
3. (See Chapter 13.) Translation of the condensate into a notation
or code.

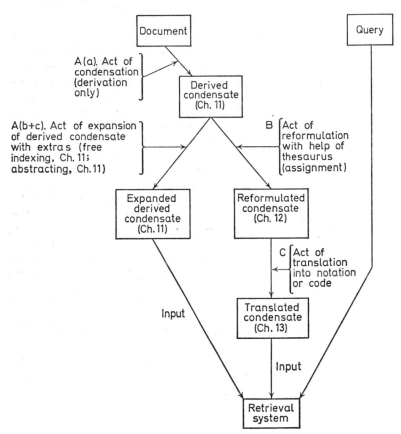

Figure 20. The link-up of documents and queries with retrieval systems

After undergoing these processes the coded information enters the
retrieval system. Short cuts of certain preparatory processes are
possible (see Chapter 14), and the choice of retrieval system is
dependent upon, among other things, such possible short cuts.

Fugmann (1970) compared the introduction of order into a certain
amount of information contained in a certain collection of documents with the second law of thermodynamics. The work of the
documentalist consists of adding energy to the system to increase
order (see *Figure 21*).

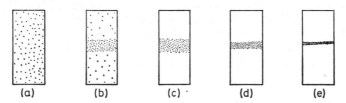

Figure 21. The introduction of order into a collection of scientific literature (Fugmann, 1970): (a) No order. (b) In indexes and cardfiles; full text storage and retrieval by computer. (c) Mechanised methods with highly developed computer-documentation languages. (d), (e) Very developed descriptor language with thesaurus and syntax, with the use of the computer. (By courtesy of the Editor of Nachrichten für Dokumentation)

It is clear that the more effort invested in this input in the system, the less effort has to be applied in the output. Or, conversely, the less effort (and money) invested in the input, the more effort (and money) must be applied in the output. The choice of the point of balance for every system depends on the set of environmental factors such as: frequency of use, urgency with which questions need to be answered, etc.

THE QUERIES

The choices in processing are influenced by the queries which the completed system would have to satisfy. In practice, there is a wide range of query situations. In general, it can be said that 'query predictability' is of importance. Factors of predictability are, for example, the level of abstraction, the degree of specialisation, the query structure, the associations that may be required, etc. Factors of predictability of queries are measured on a scale of values governed by which of the many possible practical solutions applies: (a) public library, to (b) university library, and (c) institute library; to situation (d), problem of a research team or syndicate; or even (e) collection of data in *one* project (here the boundary between documentation of literature and data is crossed).

In general, it can be said that the lower the level of abstraction and the more readily the concepts can be quantitatively expressed (data systems, for example), the greater the 'predictability' of the queries, i.e. the greater the success of the retrieval system. On the other hand, the higher the level of abstraction and the less easy the quantitative expression of the queries, the less the 'predictability', i.e. the lower the efficiency of the retrieval system. In mechanised systems a 'feedback' can be inserted whereby the system can be optimised by means of the measurement of successful handling of previous queries. Further consideration of this topic is to be found in Chapter 19. (Generally, in situations without predictability of queries no thesaurus can be made but in situations with predictability the use of a thesaurus is possible, though not necessary.)

THE USERS' AND PRODUCERS' APPROACH

Sharp (1965) summarised the problem as follows : 'The fundamental problem which is common to all is that of providing for the *nearest possible coincidence* between the *description of a subject by a searcher* and the *description used to enter documents* on that subject in the system' (the present author's italics). In other words, document content representation as it is approached by the *user* of the information storage and retrieval system must be made to correspond as closely as possible with document content representation as it is approached by the *producer* of the system.

Producers and distributors of scientific literature—the generators of the document content representation—include authors, editors, publishers, information and abstracting services and library cataloguers. Users of scientific literature include researchers, bibliographers and documentalists searching literature for other users. The two directions of approach, the producers' and the users', may be characterised by giving two typical points of difference: acceptance of complete documents as units of information and predictability of patterns of enquiry.

Macrotheme / Microtheme

The first point of difference between producers and users is the acceptance of the document as the unit of information, i.e. the book, the journal article, or whatever form in which it is published. This acceptance is normal for producers, but users tend to be interested in only a part—perhaps merely one paragraph, one subsection or even one sentence—of the published document. I suggest the terms 'microtheme' for that part of the document which interests the user and 'macrotheme' for the published document as a whole.

Document segmentation in the producers' approach

Microthemes are not used by producers of information storage and retrieval systems, because the document is usually accepted in its published form as a unit of reference. Exceptionally, the producer may use so-called links between descriptors, which indicate whether two descriptors occur together in a certain part of the document; he has then taken a step towards segmenting the document. These links are suggested as a means of avoiding noise in the search procedure. Without links or segmentation into microthemes, all descriptors are presumed to be related to the whole document, though in reality only descriptors A and B, or C and D, may be related to any particular microtheme. However, many document-alists dismiss links as superfluous sophistication (see also *Figure 28*, Chapter 15).

The microtheme as a unit of reference in the users' approach

The user often makes his own document content representation in a private file. Every research worker, documentalist or literature searcher does this document content representation at some time during or after reading professional literature. But what does he really write down for himself? A few lines, perhaps one whole paragraph, a short abstract of the paragraph or merely a few descriptors. He may also transcribe or reproduce one or two pages, but he rarely needs the whole document as published. From this activity a literature reference may ensue when he writes a paper.

A literature reference indicates to the reader either where he can find *a more elaborate discussion* on the theme or where he can find *support for the author's arguments*. The former is a typical literature reference, while the latter is more a citation. The literature reference may point to the article as a whole but often only refers to a certain part; the citation never points to the article as a whole. We shall not further consider those literature references leading to complete articles for further reading, but will confine our discussion to the majority of literature references, which are true citations. These refer to one part, a particular sentence or paragraph, i.e. what we have called a microtheme. In the literature reference ordinarily found at the end of a scientific paper, however, the whole paper referred to is always mentioned, which blurs the actual situation.

Although this way of making references referring to complete documents is common practice in the natural sciences and in technology, citations in the humanities usually refer to specific pages, perhaps because the average publication in the humanities is longer, often in book form. A citation to a book of 200 pages gives the reader little guidance; a page reference is necessary. In science and technology most research is published in short articles in periodicals and the specific page reference is usually omitted.

Every publication is constructed with microthemes as building units. If the publication is well laid out, these microthemes stand out as separate sections. Careful layout and distinct segmentation will often influence the ease of document content representation and its resultant quality (see also p. 97).

The idea of using paragraphs as microthemes in the making of document content representations has been suggested (Jacobsen, 1963; Rogers, 1964). Cleverdon (1967) uses the term 'partitioning'; Williams (1965) speaks of 'topics'. Some research on the effectiveness in retrieval of using microthemes was done by Kuhns and Montgomery (1964) but they did not obtain results indicating the optimal way to segment a document.

The anatomy of the document

The microtheme is the primary building unit of a publication. The anatomy of a publication is the design formed by the various microthemes and their mutual relationships. Occasionally authors themselves provide this design or model. For example, an article about the Dutch tomato market and the internal and external influences causing fluctuations on that market, written by Meulenberg and published in the *Netherlands Journal of Agricultural Science* in 1964, gives a diagram of its microthemes with their mutual relationships. A research worker collecting data *only* about the fluctuations of the price of tomatoes on the British market would need but a small part of this publication—one microtheme.

The history and cytology of the document

The expression 'anatomy' of the document has been used deliberately. Anatomy is a rather crude concept by comparison with the histology or cytology of the document. Anatomy in this context means the choice of indicative words which can serve as a useful representation of the content of one paragraph. Histology would mean representation of every sentence; cytology, of every word. Such a division between the anatomical, histological and cytological levels relates to the optimum depth of indexing, to which we shall return later.

Macrothesaurus / Microthesaurus

The second point of difference between the producers' approach and that of the users is the predictability of the patterns of enquiry. Producers usually have only very vague and obscure ideas about the use of content representation of the documents they produce. Their activity is discipline-oriented and the nature of the enquiries of their users is hardly predictable. However, the users know, we hope, roughly what they need. Their activity is mission-oriented and their patterns of enquiry are usually predictable. Users' needs are more predictable in retrospective searching than in reading for current awareness.

The approaches of both the producers and the users can be brought together through the use of a thesaurus. A thesaurus can be defined as a controlled vocabulary of an indexing language. The producer making such a thesaurus must have some idea of the vocabulary of the user. The more specialised the subject, the more the producer needs to adapt the vocabulary to the user group. If the producer is a commercial publisher, the number of users must be above a certain minimum; if he is a cataloguer, he can adapt his catalogue for the special use of only one or two people.

The microvocabulary of the research worker can also be called his *microthesaurus*. The microthesaurus is based on his, or his literature searcher's, pattern of enquiry. On the other hand, the *macrothesaurus* of the producer is a frame of reference which the producer deems acceptable for the user.

Users' Methods of Enquiry

How does the user present his frame of reference for studying the literature? Two situations are typical:

1. The user has a *mental picture* of the microvocabulary of the descriptors which stake out his subject. This is the normal situation: scanning the literature against this subject frame, the user chooses microthemes out of documents, which he copies by hand or by reprography.
2. What he has in mind is defined by a *written profile* which he gives to an information service for selective dissemination of information within a current awareness service, often mechanical, in which users' profiles are screened against the newly received literature. The results are indications to users of what new literature falls within their profile.

A third kind of formalisation of the frame of reference is a *written formula* of the kind of descriptors and their mutual relationships that the user is interested in. A striking example of this approach is the interesting experiment of Selye and his collaborators, who launched the SSS (Symbolic Shorthand System) (Padmanabhan and Ember, 1962). In the SSS there is a curious mixing of a descriptor language from the producers' point of view and a method of document content representation from the users' point of view. It was originally meant for general use as a descriptor language in medicine. Subsequent practical use showed that it was often successful when used by a special research group interested in specialised areas in pharmacology and endocrinology. It is thus best considered as a method of standardised document content representation for a specialised user group.

The conclusion from these remarks about the users' approach is that the user wants a very clearly articulated, well-structured presentation of the document itself; the microthemes should stand out. He is interested in the divisions of the document as it is published. The more the document content representation is built on microthemes, the better will be the precision and 'recall' which Cleverdon (1962) took as criteria for evaluating an information storage and retrieval system. This means that a document of 5 pages with 25 microthemes would appear in the information storage and retrieval system not as 1 item but as 25. But such multiplication of the number of entries by 25 is little problem for computers and their memories.

User-interest profiles are specific and indicate a high level of predictability of enquiries. The user always handles his own microvocabulary implicitly or explicitly. It matters little whether his profile and the document content representation are matched by machine or not. However, the more formal the language of both the document and the searcher's enquiry pattern, the more successful mechanised matching can be. The users' approach must be linked to the producers' approach. By comparing how they each approach document content representation, we should be able to improve our systems.

Admittedly, the foregoing implies a certain preference for methods using a thesaurus and for assignment indexing (Chapter 12), over those using derivation indexing or abstracting (Chapter 11).

The Building of the Double Bridge between Producers and Users

The construction of the necessary double bridge between producers and users in the fields of macrotheme/microtheme and macrothesaurus/microthesaurus presents no problem in only a very few particular situations. Three examples of situations in which no problems arise are:

1. The user finds no literature and makes his documents himself on the basis of experience: notes, records of experiments, etc.
2. The user finds literature and makes abstracts completely adapted to his own use.
3. A documentalist collects literature for one research worker or a *small* team working on a very special subject: here also there is a complete adaptation to the pattern of enquiries of the user(s). Even in a special library serving a medium-sized institute, this situation may still be found.

Many more situations obtain in which a problem exists in constructing the necessary bridges. The short-term solution consists in a combination of (a) an improved document presentation and (b) an integration of the production of the document with the production of the document content representation.

Improved document presentation

Document content representation may be improved by producing original documents whose logical and typographical arrangement simplify document content representation and raise its reliability and consistency with the original document. This applies to both manual and machine operations. Important elements of presentation of a publication from our standpoint can be summarised as follows.

The *lay-out* should be a clear typographical presentation of the various components of the document, indicating abstract or summary, discussion of experiments, discussion of relevant literature, conclusions, purpose of study, etc. These different parts of the document could be indicated by a system of typographical signs (Maltha, 1965). Moreover, microthemes with their descriptors should stand out, as in the typography of some school texts, where paragraphs have descriptors in the margin.

The *addition of elements* to improve document content representation might include a thesaurus (perhaps as a supplement to the general thesaurus of the journal).

It would be possible for editors of specialised journals and abstract media to produce an *editorial thesaurus* (or 'home thesaurus') of preferred terms for authors' and abstractors' use—a practice, as far as is known, not yet applied. Editors could thus control the vocabulary in certain special fields, aiding the process of standardisation of terminology and thus improving communications and certainly improving document content representation. In this way vocabulary control is shifted from the production of document content representation to the production of the document itself.

Other additional elements may include a list of suggested index-ing terms and a model of the discussion (Meulenberg, 1964).

In many cases a clearer *description of intention* in the title, captions, abstract, extract or digest is needed. Katter (1967) has studied the fidelity of these constituents by methods of pattern.

There is a need for *editorial checking* of the text, the descriptors of microthemes and the abstract against a vocabulary of preferred

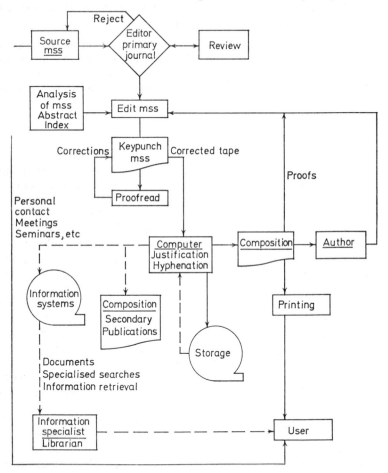

Figure 22. An integrated system of handling scientific information. In this system the central point is the computer feeding information systems as well as printing devices and secondary publications activi-ties. (From Mechanized Information Storage Retrieval and Dissemina-tion: Proceedings of the FID/IFIP Joint Conference Rome 1967, p. 50. *By courtesy of the North-Holland Publishing Co., Amsterdam)*

terms. Edited author abstracts could then be more easily lifted by abstract journals. This vocabulary should show a moderate structure for unbiased guidance of the user.

Integration of document production with document content representation

If the document of the future is printed by a computer-controlled printing process, the machine-readable original record will be available for supplementary use in documentation after the reproduction. This means an availability of all kinds of documentation devices accompanying a text including captions at the head of microthemes and glossaries, as previously mentioned (see Kuney's diagram, *Figure 22*). There is already some interesting literature on this topic of the possibility of documentation-in-source combined with computer-aided typesetting (American Institute of Physics, 1967; Vensenyi, 1968; Kuney, 1968a,b, 1970; Libaur, 1969; Lannon, 1970).

THE LITERATURE

For a general outline of the literature of the subjects discussed in this chapter, the books of Vickery (1961, 1970) and Foskett (1963) should be consulted. Mention can be made of the works of Kent (1971), Bourne (1963) and Becker and Hayes (1963), where the emphasis is, however, predominantly on 'hardware'.

References

American Institute of Physics Staff (1967). 'Techniques for Publication and Distribution of Information.' *A. R. Inf. Sci. Technol.* **2**, 339

Becker, J. and Hayes, R. M. (1963). *Information Storage and Retrieval: Tools, Elements, Theories.* New York; Wiley

Bourne, C. P. (1963). *Methods of Information Handling.* New York; Wiley

Cleverdon, C. W. (1962). *Report on the Testing and Analysis of an Investigation into the Comparative Efficiency of Indexing Systems.* College of Aeronautics, Cranfield, Bucks.

Cleverdon, C. W. (1967). 'The Efficiency of Index Languages.' In: de Reuck, A. and Knight, J. (eds.). *Communication in Science.* London; Ciba Symposium

Foskett, D. J. (1963). *Classification and Indexing in the Social Sciences.* London; Butterworths

Fugmann, R. (1970). 'Wissenschaftlichte Dokumentation und Forschungs-management.' *Nachr. Dokum.* **21**, No. 4, 160

Hines, Th. C. (1961). *The Collectanea as a Bibliographical Tool.* New Brunswick; (Unpublished Thesis) Graduate School of Library Service, Rutgers State University

Jacobsen, S. N. (1963). 'Paragraph Analysis Novel Technique for Retrieval of Portions of Documents.' *Am. Inst. Docum. 26th Annual Meeting, Chicago, 1963.* pp. 191–192

Katter, R. V. (1967). 'Measuring Fidelity in Components of Document Retrieval Systems.' Paper (mimeographed), 33rd Conference FID, Tokyo, 1967

Kent, A. (1971). *Information Analysis and Retrieval.* New York; Becker and Hayes

Kuhns, J. L. and Montgomery, C. A. (1964). 'A Comparative Study of Fragment versus Document Retrieval.' *Proc. Am. Docum. Inst.* **1**, 369

Kuney, J. H. (1968a). 'Computer Typesetting for Scientific Publications.' In: Samuelson, K. (ed.). *Mechanized Information Storage Retrieval and Dissemination: Proceedings of the FID/IFIP Joint Conference, Rome, 1967.* p. 510. Amsterdam; North-Holland

Kuney, J. H. (1968b). 'Publication and Distribution of Information.' *A. R. Inf. Sci. Technol.* **3**, 31

Kuney, J. H. (1970). 'New Developments in Primary Journal Publication.' *J. Chem. Doc.* **10**, 1, 43

Lannon, E. R. (1970). 'Subject Indexing as a Byproduct of Electronic Composition.' *J. Chem. Doc.* **10**, No. 1, 46

Libaur, F. B. (1969). 'A New Generalized Model for Information Transfer. A Systems Approach.' *Am. Docum.* **20**, 381

Maltha, D. J. (1965). *Symbols for the Designation of Different Parts of Articles.* Wageningen; Centrum voor Landsbouwdocumentatie

Meulenberg, M. T. G. (1964). 'A Quantitative Investigation into the Dutch Tomato Market: A Seasonal Analysis.' *Neth. J. Agric. Sci.* **12**, 169

Padmanabhan, N. and Ember, G. (1962). 'Symbolic Shorthand System for Physiology and Medicine.' *Meth. Inform. Med.* **1**, 138

Rogers, F. B. (1964). 'The Relation of Library Catalogs to Abstracting and Indexing Services.' *Libr. Q.* **34**, No. 1, 106

Sharp, J. R. (1965). *Some Fundamentals of Information Retrieval.* London; Deutsch

Vensenyi, P. E. (1968). 'Indexing in Source.' *Coll. Res. Libr.* **29**, No. 5, 400

Vickery, B. C. (1961). *On Retrieval System Theory.* London; Butterworths (2nd edn, 1965)

Vickery, B. C. (1970). *Techniques of Information Retrieval.* London; Butterworths

Williams, W. F. (1965). *Principles of Automated Information Retrieval.* Elmhurst, Ill.; The Business Press

ELEVEN

PROCESSING OF DOCUMENTS PRIOR TO ENTRY INTO THE RETRIEVAL SYSTEM
(Methods of condensation with derivation)

The success of all information storage and retrieval systems depends on document content representation. 'The quality of information storage and retrieval systems can only be as good as the quality of document content representation,' says Taulbee (1968). Stevens (1965), in his discussion of the producers' approach, classified the methods of document content representation into those using *derivation* and those using *assignment*. Derivation takes descriptors from the text without vocabulary control; assignment takes descriptors from an established list so that vocabulary is controlled. Two further methods may be considered: *free indexing*, which allows derivation of descriptors together with the use of any other words of natural language; and *abstracting*, which usually combines both derivation and a sort of assignment based on precepts.

In Chapter 12 the methods with *assignment* will be discussed. In this chapter attention will be given to the three other methods in the following order: (1) derivation, (2) free indexing, (3) abstracting.

METHODS WITH DERIVATION

The optimum condensation of a document (or part thereof) depends upon the query situation and can lie between a maximum of all the words of the document and a minimum of a single word (subject heading). The smaller the condensation, the smaller the loss of information; the greater the condensation, the greater the loss.

The three methods of document condensation with derivation can be grouped as follows:

1. Condensation of each sentence severally, taking *all* sentences into account.
   (a) *All* words, except those on a list of exceptions.
   (b) Meta-language.
2. Condensation by selection of only certain sentences, which remain integral.
3. Condensation *without* preservation of each sentence or of the sentence as such. Statistical word frequency analysis.

A review of all these methods, in so far as they are mechanised, can be found in Krallman (1968).

Condensation of Each Sentence Severally, Taking All Sentences into Account

All words, except those on a list of exceptions

An example of this system is that described by Simmons and McConlogue (1963): a mechanical analysis of each sentence is undertaken and a VAPS number is allocated (V=volume, A=article or chapter, P=paragraph, S=sentence), giving each sentence its definite retrievable place or number. All words are taken up, with the exception of those on a special list. This method of working borders on the total storage of the document (see Chapter 14).

Meta-language

All attempts to form a 'meta- or inter-language' aim at a certain sentence standardisation. In this case the natural 'redundancy' and the indistinct concept linkage of normal speech would be avoided.

De Grolier (1962) gives a survey of these experiments under the title 'Techniques aimed at using a more or less standardised form of natural language.' For example, sentences can be standardised into a small number of structure types (Harris, 1959). To these experiments belongs the pioneer work of Gardin (1960a,b) in the field of the humanities. Gardin outlined methods for 'linguistic reduction'. In this sub-section mention must be made of the theoretical discussion of workers in structural linguistics whose investigations were undertaken not for documentation but for machine translation research, and many of whose outstanding experiments are reported by de Grolier (1962). See (1964) presents a survey of experiments in the field of machine translation, where he distinguishes five different levels of increasing complexity:

1. Item-for-item substitution.
2. Morphological processing, which takes into account the morphology of the language.
3. Syntactic translation, based on the syntax of languages.
4. Transformational translation, with transformation of the original syntactic construction to a kind of normalised simpler construction.
5. Semantic level.

One important difference between machine translation and condensation for documentation is: translation activity accepts the original text as it is and the carriers of the message (words or notations) are translated. In condensation for documentation, however, it is not the carriers which are translated but the message itself. In the first case there is no selection or evaluation, but both are present in the second case.

For additional literature reference can be made to Delavenay (1960) and Kent (1960), for example. Periodicals which cover developments in the field of mechanical translation include *Bulletin trimestriel de l'Association pour l'étude et le developpement de la traduction automatique et de la linguistique appliquée* (ATALA) and *Information storage and retrieval including mechanical translation*. The point at issue is whether standardisation of natural languages is at all possible. Harris (1959) is an example of a research worker who affirms these possibilities. De Grolier (1962) is sceptical; Beier (1960) and Haldenwanger (1961) point out the impossibility of standardising a living language without such a loss of information that standardisation would serve no useful purpose. The opposite poles of written language are, *on the one hand*, formula and sign languages (as in mathematics, chemistry and chess), in which everything is committed to the ciphers and nothing is left to the reader's imagination, and, *on the other*, literary languages (poems!), in which so much is read *between* the lines and can only be conjectured. The first group is already standardised; the second can never be so. (See also Chapter 14, p. 30.)

Gardin (1960a) warns against too much standardisation, which would cause input to become too expensive while reducing output costs; with too little standardisation the reverse is achieved.

Salton (1968) has published a comprehensive book about what can be regarded as the most elaborate experiments in this field. His SMART system represents an ingenious combination of techniques of reduction, frequency-counts, inclusive co-occurrence of words, statistical phrases, syntactic trees, etc. The final unit underlying the SMART system is the word-string.

Condensation by Selection of Only Certain Sentences which Remain Integral

Here is the field of autoabstracting of Luhn (1958), in which the choice of sentences depends on the statistically determined (by word frequency and vicinity statistics) importance of the sentences. This statistical method of finding '*phrases-cléfs*', as the French research workers Braffort and Leroy (1959) call them, has met with considerable criticism: it is just as difficult to standardise a living language satisfactorily as it is to analyse it statistically. It is worth while mentioning that Clark (1960), for example, has established that in the field of anthropology (description of certain rituals among primitive peoples) Luhn's method has been successful. Clark states that the method has shown itself to be one of 'astonishing ingenuity'. Climenson, Jacobsen and Hardwick (1961) seek a diagnosis of sentence structure by machine means; the sentences are then divided into two groups with 'central and peripheral structure'. Sentences with a 'central structure' are then selected, and the theory is that with this selection a document can be condensed to 35 per cent of its original length with negligible loss of information. The selection of sentences may be carried out by application of the following methods.

Statistical methods use a direct deduction from the statistics of word occurrence or of co-occurrence of word pairs. A machine selects sentences by the frequency of occurrence of words within sentences of the text. Sentences with four or more high-frequency words would be preferred to sentences with a lesser number.

With *linguistic methods* the machine selects sentences of a certain syntactic structure (e.g. those beginning with a subject modified by attributives) or sentences with particular expressions in the text (e.g. those containing 'conclusion is ...'). These structures and/or expressions must be programmed in advance.

Methods based on *textual form* include, for instance, a special place in the text (headings, captions) and special typographical details (indentation, bold face, italics).

Much research has been done on the selection of sentences, but Sharp (1967) reports that the results so far are disappointing.

These methods of choosing sentences have been called 'automatic abstracting', but it is clear that they are really 'automatic *ex*tracting'. We will use the term 'abstracting' later in the discussion, in cases where a document content representation involves a critical choice.

One extreme example is the selection of only one sentence, namely that of the title: this occurs in the K(ey) W(ord) I(n) C(ontext) system, as employed, for example, in *Chemical Titles.*

In 'current contents' methods the title is likewise the only part derived from the publication. But the difference is that in current contents situations the titles of the papers are reproduced as they appear in the contents list of the journal issue. The KWIC index provides the user with additional searching aids by permuting the title for all possible search words contained in it. Fischer (1966) gives a review of the literature on KWIC indexes.

Condensation without Preservation of Each Sentence or of the Sentence as Such

This condensation is semantic condensation; the sentence as such no longer plays any part. The treatment outlined in the previous two sub-sections can be said to be syntactic condensation.

Statistical word frequency analysis

Word frequency in itself is not a norm for importance. On the other hand, a high level of condensation can be achieved by enumerating distinct words: Vickery (1961b) has estimated that a book of nearly 58 000 words can be condensed to 7 per cent of its volume by listing 'each distinct word without paying attention to its frequency'. Vickery (1961a) has also shown that the word frequency method has a great measure of success in so far as: (a) homogeneity of text, (b) compilation and use of an extended list of excluded terms of low information value.

Meyer (1961) is very critical and compares the results of word frequency methods with those of his suggested document analysis by means of concept chains. Sharp (1967) states that recent research has made us sceptical about the possibilities of success of this method, and Werzig (1970) gives a long list of points against text analysis by word frequency counts.

METHODS WITH FREE INDEXING

The methods of free indexing are free from both original texts and rigid prescriptions. However, most of these free indexing methods include some kind of prescription, though only of a very general nature, e.g. how many descriptors to use, whether to relate the descriptors syntactically or to use certain word forms (e.g. verb or substantive; substantive or adjective; plural or singular). A good example of such prescriptions is the guidelines edited by COSATI (1967). Although the choice of the descriptors is free, most of them do occur in the document anyhow. Thus the result is a mixture of derivation and free choice. It is also called 'enriched indexing', because it enriches the original vocabulary of the document. Incidentally, the index for one independent book is usually made by free indexing, as consistency is no problem from the producers' approach.

Free indexing was originally used in indexing by Uniterms when the choice of Uniterms was unrestricted. The principle of the Uniterm is that by reducing subject headings to single terms, the total number of terms is reduced because of repetition of words within subject headings. It is possible to go further and make the Uniterm a word root (e.g. ICE; all other terms—iced up, icy, iced over—being subsumed to this root) keeping the list even shorter (Taube, 1953). As the number of indexed documents grows, however, the same Uniterm will apply to a larger number of documents, and the number of Uniterms required in searching will be far higher than the more specific subject headings in a similar situation.

Bourne (1963) distinguishes a series of possibilities arranged according to the principle of *ascending structuralisation*. He gives the following levels:

1. Words chosen from title or text but common words omitted.
2. Words chosen from title or text but common words omitted and consideration of variants.
3. Words chosen from title or text but common words omitted and consideration of variants and generic relationship.
4. Words chosen from title or text and consideration of syntactical relations between indexing terms.
5. Any of the preceding methods with additions of terms *not used in the text* (this step requires judgement of subject knowledge).
6/7. Fixed authority lists (see Chapter 12).

METHODS WITH ABSTRACTING

Abstracting largely draws upon words used in the document itself. In this sense abstracting is a method of derivation. However, abstracts journals always supply their abstractors with stringent instructions. There is an element of assignment in this. There is no fixed list of descriptors, but there is a fixed technique of document content representation: e.g. to include conclusions and results, to mention methods, to exclude descriptions of experiments and discussion of the literature. Such a practice is a normal feature in the larger abstracting services.

Two problems are especially relevant to document content representation:

1. Is an author abstract preferable to a non-author abstract? The author abstract is more economical, but should be checked by the editor of the journal in which it will be published. This should remove the idiosyncracies of an author abstract. The editor may even check the words used in author abstracts against a vocabulary of preferred terms.
2. For non-author abstracts Borko and Chatman (1963) found that instructions of abstracting journals are nearly all alike. They have devoted an extensive study to these instructions, which reveals that the instructions display a greater similarity than might have been supposed; in the final analysis all emanate from a small number of minimum criteria.

How the form of presentation of one and the same publication in different abstracts varies and how great is the dependence upon the abstractor have been shown by MacMillan and Welt (1961). How up to five different journals may abstract the same chemical paper differently was described by Schüller (1960). The five different abstracts can be found in Weil, Zarember and Owen (1963). No the abstracts may be caused by unclear structure of the original and by different disciplines of the abstractors (e.g. biochemistry, physical chemistry).

A review of the discussions on the extent of information loss in abstracts can be found in Weil, Zarember and Owen (1963). No. unanimous opinion as to the amount of loss has been agreed upon, but there is a general warning to reduce reading speed considerably in the case of abstracts, because of the high concentration of information.

Finally we mention analysis by means of concept chains as evolved by Meyer (1961), with special reference to documentation of patents. By this method considerably *more* is covered than is covered by the abstract method but much less than by a 'metalanguage'. (For the abstracts journal, see Chapter 4, p. 31.)

References

Beier, E. (1960). *Wege und Grenzen der Sprachnormung in der Technik.* Dissertation, Bonn

Borko, H. and Chatman, S. (1963). 'Criteria for Acceptable Abstracts: a Survey of Abstractor-instructions.' *Am. Docum.* **14**, No. 4, 149

Bourne, C. P. (1963). *Methods of Information Handling.* New York; Wiley

Braffort, P. and Leroy, A. (1959). 'Des mots cléfs aux phrases-cléfs.' *Bull. Biblioth. Fr.* **11**, 383

Clark, L. L. (1960). 'Some Computor Techniques in the Behavioral Sciences.' In: Kent, A. (ed.). *Information Retrieval and Machine Translation.* Vol. 1, Ch. 11, pp. 445–466. New York; Interscience

Climenson, W. D., Jacobsen, S. N. and Hardwick, N. H. (1961). 'Automatic Syntax Analysis in Machine Indexing and Abstracting.' *Am. Docum.* **12**, 178

Delavenay, E. K. (1960). *Bibliography of Mechanized Translation.* The Hague; Muutain

Fischer, M. (1966). 'The KWIC Index Concept: A Retrospective View.' *Am. Docum.* **17**, No. 2, 57–70

Gardin, J. C. (1960a). 'Document Analysis and Information Retrieval.' *UNESCO Bull. Libr.* **14**, No. 1, 2

Gardin, J. C. (1960b). 'Etudes séminologiques et documentaires. Recherches en cours.' *Bull. Biblioth. Fr.* **5**, No. 11, 443

Grolier, E. de (1962). *A Study of General Categories Applicable to Classification and Coding in Documentation.* Paris; UNESCO

Guidelines (1967). *Guidelines for the Development of Information Retrieval Thesauri.* COSATI

Haldenwanger, H. M. (1961). 'Begriff und Sprache in der Dokumentation.' *Nachr. Dokum.* **12**, No. 2, 65

Harris, Z. S. (1959). 'Linguistic Transformation for Information Retrieval.' *Proc. Int. Conf. Scient. Inf. Washington*

Kent, A. (Ed.) (1960). *Information Retrieval and Machine Translation.* 1960. New York; Interscience

Krallman, D. (1968). 'Maschinelle Analyse natürlicher Sprachen. Part 4: Automatische Dokumentation.' In: Gunzenhäuser, R. (ed.). *Nicht numerische Informationsverarbeitung.* Berlin; Springer

Luhn, H. P. (1958). 'The Automatic Creation of Literature Abstracts.' *J. Res. Dev.* **2**, No. 2, 159

MacMillan, J. T. and Welt, I. D. (1961). 'A Study of Indexing Procedures in a Limited Area of the Medical Sciences.' *Am. Docum.* **12**, No. 1, 27

Meyer, E. (1961). *Grundfragen der Patentdokumentation.* 2nd edn. 1961. Düsseldorf; Vlg. V.D.I.

Salton, G. (1968). *Automatic Information Organization and Retrieval.* New York; McGraw-Hill

Schüller, J. A. (1960). 'Experience with Indexing and Retrieving by UDC and Uniterm.' *ASLIB Proc.* **12**, No. 11, 372–89

See, R. (1964). 'Mechanical Translation and Related Language Research.' *Science, N.Y.* **11**, 621

Sharp, J. R. (1967). Content Analysis, Specification and Control.' *A. R. Inf. Sci. Technol.* **2**, 87

Simmons, R. F. and McConlogue, K. L. (1963). 'Maximum Depth Indexing for Computer Retrieval of English Language Data.' *Am. Docum.* **14**, No. 1, 68

Stevens, M. E. (1965). *Automatic Indexing: a State-of-the-Art-Report.* Washington; National Bureau of Standards (NBS Monograph 91)

Taube, M. *et al.* (1953). *Studies in Coordinate Indexing.* Vol. 1. Washington; Documentation Inc.

Taulbee, O. E. (1968). 'Content Analysis, Specification and Control.'*A. R. Inf. Sci. Technol.* **3**, 105

Vickery, B. C. (1961a). 'The Statistical Method in Indexing.' *Rev. Int. Docum.* **28**, No. 2, 56

Vickery, B. C. (1961b). *On Retrieval System Theory.* London; Butterworths (2nd edn, 1965)

Weil, B. H., Zarember, I. and Owen, H. (1963). 'Technical Abstracting.' *J. Chem. Docum.* **3**, No. 2, 86

Werzig, G. (1970). In: Schiber, H. W. (ed.) *Dokumentanalyse als sprachlich-information Theoretische Probleme.* Beitr. Inf. Dok. Wiss. Folge 2

TWELVE

PROCESSING OF DOCUMENTS PRIOR TO ENTRY INTO THE RETRIEVAL SYSTEM
(Methods of condensation with assignment)

INTRODUCTION

With all *methods of derivation* searching is more cumbersome than with methods of assignment. The lack of vocabulary control leaves the user uncertain in many situations.

In *methods of assignment* the words for document content representation are derived from a prepared vocabulary of descriptors, i.e. a *thesaurus* or an authority list.

The term 'descriptor language' is used for all kinds of thesaurus, and for classifications as well. The word 'thesaurus' has many definitions in the professional literature, as Vickery (1960b) has pointed out. Our use of the word has already been defined in Chapter 10, p. 95.

The thesaurus is an aid which is used preferably when the predictability of the query is great enough to permit limitation to certain words of the language. This situation arises in practice where there is a distinct field, a complex of problems or even a single problem.

What is the exact function of this list? It is the bridge for the research worker from his own terminology to that of the list, i.e. a kind of 'standardisation'. The chosen words (descriptors) can only be used to the exclusion of all other descriptors. It should be noted in addition that the use of cross-references (from singular to plural, from one synonym to another) excludes words from the list. This exclusion is based merely on grammatical norms and does not reach the stage of a thesaurus, which is reached only when semantic (conceptual) norms are employed as factors of selection.

Thus it is a matter of degree in 'control of vocabulary', i.e. merely a quantitative difference, and therefore a matter of agreement as to when a thesaurus actually exists. The situation is clearest in U(niversal) D(ecimal) C(lassification): the subject index in UDC is not a thesaurus; the systematic list of UDC numbers is, but the descriptors are numbers instead of words.

INTERNAL ORGANISATION OF A THESAURUS

One of the main problems is the *internal organisation* of the thesaurus. If it is short, there are no pressing problems; but the longer the thesaurus, the more structure becomes desirable.

The first elements of internal structure are the *see* and *see also* references; more elaborate are the thesauri including NT (narrower term), BT (broader term), and RT (related term), which constitute a signal pattern leading the searcher along a more complicated search path.

An example of the introduction of more structure is facet classification, and the ultimate organisation is the well-known hierarchical classification, as exemplified by UDC with its primary dependence on general to specific relationships.

At the 1964 Elsinore Conference of classification expert theorists defined the term 'classification' in this way: 'By classification is meant any method creating relations, generic or other, between individual semantic units, regardless of the degree of hierarchy contained in the systems and of whether these systems would be applied in connection with traditional or more or less mechanized methods of document searching' (Atherton, 1965, p. 544). Note: *any* method of creating relations, which means any application of structure in a list of chosen terms.

We cannot say whether the Elsinore formulation will bring lasting peace from the ancient battle of classifications versus alphabetically arranged subject headings. It does, however, point to the great range of possible descriptor languages, from the very informal to those with a rigid hierarchical structure. Hopefully, the Elsinore definition should end the polemics.

Jonker (1959) speaks in this context of a 'descriptive continuum'.

It is still difficult to concentrate on the *similarities* of all descriptor languages, unstructured, less structured or fully structured, since there is a long tradition of antagonism between alphabetical subject headings on the one side and classification on the other. It is therefore important to bear in mind (a) the sliding scale of the levels of structure; (b) the fact that any classification is accompanied by an alphabetical index and that every thesaurus of any volume has its counterpart (usually unpublished but in the possession of the controlling team) in a classified list of the descriptors used in the thesaurus. (This list is a necessary tool at the moment of consideration of new descriptors for inclusion in the thesaurus.)

CONSTRUCTING A THESAURUS

Three problems arise:

1. How to Select Descriptors for the Thesaurus?

Published lists of subject headings may be used as a source as well as existing classifications. If such lists can be used with a few modifications, this is the quickest and cheapest way. The best, but certainly the costliest, way is to compile a thesaurus tailored to fit the job in hand. This compilation was often undertaken in the past theoretically—that is, in a classification of the particular field. The modern tendency is to select a series of documents representative of the collection under review which are to provide elements for a list in the first instance ('literary warrant'). Statistical methods are applicable to this selection, and with mechanical means for word frequency counts of single words, word pairs and so on, already familiar in documentary analysis, a practical guide to the construction of a list is to hand. This method of thesaurus building is called automatic indexing. Batty (1969) and Stevens (1965) have written review articles on this subject. However, this same term 'automatic indexing' is also used for another activity.

The full text of the document is processed in the machine, which has been programmed with the thesaurus. The result of matching text with thesaurus is a list of indexing terms suitable for retrieving the document. This method of automatic indexing based on a prepared vocabulary is one with automatic input and differs from the method of automatic indexing previously mentioned for thesaurus building (see Borko, 1968, and Lustig, 1969).

A special problem is the choice of the level of a descriptor. To what extent must it be split into composite concepts? Kent (1971) partitions the concept 'thermometer' in three semantic factors, machine ... measuring ... temperature, and the concept 'barometer' in machine ... measuring ... air pressure. The same problem arises with chemical compounds: the composite compound name or the constituent elements? The problems have been worked out by Foskett (1961) and the Classification Research Group in the discussions about the so-called 'integrative levels'. The mini-vocabularies (Moss, 1967) which are generated in that way are easier to handle than larger ones but input and output in the selection devices become more onerous and thus more expensive. The opposite of semantic factoring is semantic integration in new words covering complicated composite concepts. On the other hand, these compositions reduce the possible permutations of the composing elements (see also Chapter 15, p. 139).

2. How Much Structure should be Built in?

If, in the evaluation of the actual query situation, the predictability is deemed great enough to justify a method of document analysis by means of a thesaurus, the next problem is the amount of built-in structure, and this depends in turn upon the expected structure of the queries and how much guidance would be required to terms of supra-, infra-, or juxtaposition. In general, this problem of structure is also closely linked with the particular field of knowledge concerned.

Structure in a vocabulary is a helpful device for searchers. The help is virtually proportional to the amount of structure introduced into the vocabulary. A hierarchical classification gives most help (though perhaps a biased help); facet classification, less; and reference to broader, narrower and related terms, still less. The introduction of syntax (e.g. role indicators) is certainly a helpful device for searchers. More structure means more help and less need for initiative in search strategy; less structure means less help in searching and more need for initiative, but also more freedom in formulating a search strategy.

3. How is Structure Built in?

Associative patterns

In associative patterns structure may be built in by *see* and *see also* references. As already stated, BT, RT and NT are often used in thesauri. With regard to the application of mechanical means in this field, the following can be said.

In constructing a list by means of frequency counts an important starting point is the strength of association between two terms and its measurability. This depends upon searching a representative set of documents for term affinities and word 'clumps' which can be isolated. Work is carried out in this field by means of a type of experimental word association coefficient determined statistically. This results in 'approximation tables' or even atlases (or indeed such other graphic representation as the circle used by Schüller, 1964) which show the proximity values of terms. In this connection the works of Doyle (1962), Needham and Sparck Jones (1964) and Lancaster (1964) can be cited. Rolling (1965) discusses the various diagrammatic representations. The great danger is the starting point: an existing collection of documents tends to 'freeze' into a certain pattern and thereby hinder future new developments.

Hierarchical patterns

Figure 23 gives a simple example of building-in without statistical aids a structure in an originally unstructured subject heading list (first column) compiled from a series of documents and then used as a thesaurus. Here the transition is from a purely alphabetical subject heading list by way of a faceted classification to a mono-hierarchy with a built-in (arbitrarily) hierarchy of criteria of division. A faceted classification is a polyhierarchic classification, in which each small hierarchy, however, is enumerated by *one* single criterion of division. (Thus, e.g., 'Chairs' either by *material* or *style* (Groeneveld, 1954).)

Subject headings and their abbreviations	Purely alphabetical subject heading list	Structuring but now so that each structure (classification) shows only one criterion of division (added in parentheses) (polyhierarchy, faceted classification)	The same but now monohierarchical without regard for type of criteria of division
Danish milch cattle DM	CD	DM	
Milch cattle MC	CG	MC < (geographic)	
Netherlands milch cattle NM	DM	NM	DM
	MC	MC	MC<
Cattle, general CG	MM	CG < (utility)	CG< NM
Cattle, draught CD	NM	CD	MM< CD
Pigs PG	PG	PG	PG
Mammals MM		MM < (zoologically CG systematic)	
Number of independent subject headings	7	3	1
Necessary cross-references	0	6	6

Figure 23. Building-in structure

THE MONOHIERARCHIC CLASSIFICATION

The debate about the choice of the amount of structure to be built in is still going on. The older monohierarchic classification is steadily losing ground; its disadvantages can be named as:

1. Lack of flexibility.
2. Hierarchy of the characteristics of division is built in.
3. Within a monohierarchy a descriptor has a direct relation to one (and one only) broader term.

With regard to (1), Ariès (1957), a confirmed opponent of monohierarchic classification, states:

> Our experience over ten years has also shown the difficulties of a classification framework predetermined and systematically constrained; these difficulties grow with the specialization of the documents and analyses. As a result, since 1956 we have eschewed the all-too-rigid formula of classification and therefore each document has been classified by the subject headings explicitly given by the document itself disregarding any existing framework.

De Grolier (1962, Chapter 11) says that UDC is too hierarchic to be really viable; however, since it has become an 'institution', it will continue to exist 'but the future lies elsewhere'.

With regard to (2), a monohierarchic classification (UDC, for example), implies a judgement of value on the criteria of division. The criterion deemed to be most important takes precedence, the others following in the order of their respective values. So long as the problem directed to the system shares the same judgement of value (and therefore the same succession) of criteria of division, the system works. But if a problem with a secondary or tertiary criterion of division is raised to the primary position, the system fails.

With regard to (3), *Figure 23* clearly shows the difference between monohierarchic classification and the two other possibilities. Only in an elaborate alphabetical index on such a pattern can the relations to other broader terms be mentioned.

A classification with varying criteria of division on differing planes is called heterogeneous classification by Groeneveld (1954). He gives as an example the possibilities of division of furniture, e.g. according to style, material and function. In a monohierarchic classification the decision is taken to follow a certain order, i.e. hierarchy of criteria of division (for example, first by style, then by function, and finally by material).

All this does not mean in any way that the role of the mono-hierarchic classification has ended. The tasks that can be undertaken by monohierarchic classifications are as follows:

(a) Groups of general terms, recurring in all subdivisions, can be made into a general and universally applicable arrangement (and in most cases a corresponding system of significant codification): for example, headings of time, place, etc.

(b) The major classification systems might take on the supervision of usages of notation and issue recommendations as to the most appropriate practices.

Naturally tensions arise within the organisations dealing with general classifications because of modern developments. In such organisations two opposing tendencies are to be observed, namely:

(a) The tendency of users of large major systems and libraries to lay great stress on the availability of 'pigeon-holes' in the various fields. They are, so to speak, 'centripetally' orientated, deductive. They are interested in the main divisions and gladly avail themselves of monohierarchic classification.

(b) The tendency of the specialist users, who find universal classifications increasingly difficult to manipulate, to seek their own, inductively constructed systems instead of main divisions and a monohierarchic classification. The users of these specialised systems could be designated 'centrifugal'.

THE POLYHIERARCHIC CLASSIFICATION

The polyhierarchic classification is a scheme of division into categories. An example of this is faceted classification, originally used by Ranganathan with a few constantly recurring facets. Vickery (1959) has modified this classification somewhat, arriving at an expansion of the number of facets (categories).

Faceted classification is an example of a homogeneous classification, i.e. of a classification in which each facet is based on only one criterion of division. Keeping to the previous example: there would be one facet according to *function*, one according to *style*, and one according to *material*.

Historically, facet classification came last (*Figure 24*). Botany serves as an example: 'and whatsoever Adam called every living creature, that was the name thereof'. Adam made an enumerative list (*A* in *Figure 24*). Linnaeus evolved a major classification of all

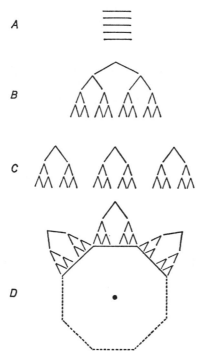

Figure 24. The development of knowledge from the stage of mere enumeration to the polyhierarchic stage

plants (*B* in *Figure 24*). Modern botany is interested in the structure of the subdivisions (*C* and *D* in *Figure 24*), e.g. ecology, taxonomy, physiology and so on, but less interested in the connections of these subdivisions to one single monohierarchic classification, especially since these higher links are debatable. Thus a polyhierarchy develops (*C* and *D*), *D* showing how the several small hierarchies are directed to one and the same object, i.e. the plant. It can be said in this outline that Stage *A* of enumeration corresponds to an alphabetical list; Stage *B* corresponds to a monohierarchic classification; and Stages *C* and *D* are polyhierarchic. The botanical parallel is true only to a certain degree: Vickery (1962) says that classification in taxonomy is fundamentally different from classification in documentation. He describes the latter in the following way: 'Documentary classification is a condensed standardised vocabulary with a simple standardised syntax possibly related to natural language by a set of transformation rules.'

COMPARISON OF ASSOCIATIVE AND HIERARCHICAL PATTERNS

Figure 25 may be used to compare monohierarchy and facet classification with a non-hierarchical pattern based on associations between descriptors. It is constructed on the basis of the relations of the phrase 'soil types' in three different patterns: (a) associative pattern; (b) facet classification; (c) monohierarchic classification.

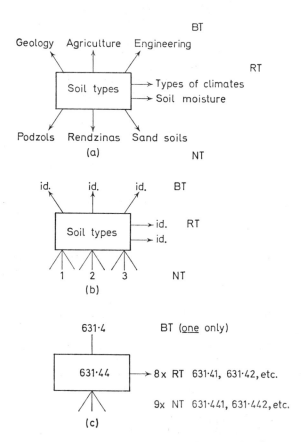

Figure 25. Comparison of associative and hierarchical patterns. (a) Associative pattern. (b) Facet classification: 1, division by particle sizes; 2, division by colour; 3, division by moisture content. (c) Monohierarchic classification—division not by one criterion only

EVALUATION OF DESCRIPTOR LANGUAGES

Cleverdon has done important research on the evaluation of descriptor languages. It is both surprising and reassuring that he found no essential differences in success in retrieval as between various thesaurus and classification systems (Cleverdon, 1967). The main result was the development of two criteria for measuring performance. These criteria are: recall and precision. Recall is the number of relevant documents retrieved expressed as a percentage of the total number of relevant documents present in the collections. Precision is the number of relevant documents expressed as a percentage of the total number of retrieved documents, including the extras (noise) (Cleverdon 1962; Cleverdon and Mills, 1963).

Vickery (1963) has clearly explained the scientific viewpoint by a very clear and simple scheme (*Figure 26*).

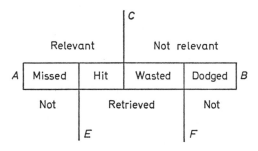

Figure 26. Recall and relevance (after Vickery, 1963, by courtesy of Journal of Documentation)

AC should be covered as far as possible by *EF*, whereby *AE* and *CF* should be as small as possible. It can be said that the nearer a retrieval system approaches this ideal, the better it is from the scientific point of view.

Cleverdon uses the measures *recall* (translated in Vickery's illustration as *EC/AC*) and *relevance* (later, precision; Vickery, *EC/EF*) as criteria of evaluation.

Bornstein (1961) indicates that there should be a method to establish when a document may be deemed relevant in reply to a query and raises the question that there might well be degrees of relevance which have not yet been measured. In addition, factors which may have a bearing on improvements in recall and precision are now under discussion.

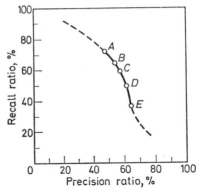

Figure 27. Recall/precision relation, from Lancaster, F. W. (1968) Information Retrieval Systems, p. 152. (*By courtesy of J. Wiley & Sons Inc., New York*)

In *Figure 27* Lancaster gives an example of the relation between recall and precision. *A, B, C, D* and *E* represent the recall/precision performance with an increasingly refined search strategy with more and narrower descriptors. Lancaster's comments are: 'The most precise strategy (*E*) will retrieve fewest documents and presumably achieve the highest precision, but the lowest recall. The less precise strategy (*D*) will retrieve more documents, improve recall, but reduce precision, and similarly for *C* and *B*.'

In *Information Retrieval Systems* Lancaster enumerates the following aids to increase recall:

'1. Enlarge the size of document classes.
2. Broaden term definition.
3. Reduce vocabulary size.
4. Synonym and quasi-synonym control.
5. Word form control.
6. Hierarchical linkage.
7. Clumping and clustering.'

He also enumerates seven aids to increase precision:

'1. Reduce the size of document classes.
2. Restrict term definition.
3. Increase vocabulary size.
4. Coordination.
5. Weighting.
6. Interlocking.
7. Role indicators.'

One of the main points arising from the Cranfield experiment was the exact description of 'relevance', which proved to be a difficult criterion depending on perfectly subjective judgements. Cuadra and Katter (1967) have discussed this black box of relevance. A recent monograph is that of Gilchrist (1971).

THE LITERATURE ON THESAURI AND CLASSIFICATION

Blagden (1968) has compiled a literature review on thesaurus building. Guidelines for thesaurus building in the framework of the studies for 'Unisist' are reported by Lorch (1970). Soergel (1969) has studied classification systems and thesauri. Practical examples in specific fields are given by Rahmann and Samulowitz (1968) and Scheele (1967). The review of Fairthorne (1970) also gives various examples. Mandersloot, Douglas and Spicer (1970) discuss devices for control of a thesaurus. Baulkwill and Posnett (1969) describe their method of derivation of a thesaurus from a classification.

A number of references on classification may be added for those who wish to study the subject in greater depth.

The very valuable manual by Vickery (1959) acts as a signpost to the development of ideas on classification. It is most suitable as an introduction to private study of classification. The works of Coates (1960) and Foskett (1963), too, are illuminating and provide good guides to the complex of problems discussed in this chapter.

For the historical development of classification systems, reference can be made to Lamontagne (1953) and Shera (1951). The articles by Holmstrom (1947), Foskett (1962) and de Grolier (1956) should also be mentioned. The tendency to polyhierarchy occurs as a 'leitmotiv' in various writings on classification. Broadfield (1946) speaks of 'the Utopian character of a general classification of sciences' and is of the opinion that each specialist classification can only survey the question from one point of view. Egan (1953) writes that for a long time universal bibliographies have been recognised as useless white elephants; the corollary question, however, apparently still remains whether or not universal classifications can be replaced. Egan says that it might be possible to cluster from a number of words one group, representing a single viewpoint; that is to say, purely inductive. Taube (1951) with his 'bibliographic control' merely means a number of independent viewpoints in a special, limited field. Brisch (1955) calls his independent points of view 'chapeau-headings'; his system was warmly reviewed by Kyle (1956). Mooers (1956) was possibly the first to coin the expression 'descriptors'. Bohnert (1955) gives juxtaposed examples of various polyhierarchic systems, describing their practical application.

Vickery (1951) goes further into the lack of a clear international nomenclature and sees in the 'fundamental constituents' of Ranganathan, the 'isolates' of Farradane, the 'descriptors' of Mooers and his own 'semantic elements' only different terms for one and the same thing. The same author has written an important survey (1955) of the parallel development of facet analysis (Ranganathan), Uniterm co-ordinate indexing (Taube) and the use of subject headings, in which he lays special emphasis on the far-reaching measure of agreement.

In London, on the initiative of the Scientific Information Conference of the Royal Society, a Classification Research Group has been at work since 1948. Their proceedings appear in the bulletin published by this Group (later transferred to *Journal of Documentation*) and the Proceedings of the Dorking Conference (1957) and of the Elsinore Conference (1965), edited by Atherton.

The conservative, tentative conclusions of CRG state that each system requires a classification as a basis and that the only question would be as to what kind of classification would result. The opinions of the participants tend towards faceted classification.

Faceted classification received its name from Ranganathan (1951). Because of his involved style, other literature on 'faceted analysis' is to be recommended, e.g. Vickery (1954), Palmer (1953) and Palmer and Wells (1951). A survey of the literature is to be found in Chatterji (1952). Farradane (1950, 1952) has also published important theoretical articles. The 'know-how' of faceted classification can be found in Vickery (1960a). In particular, many references are collected in a survey study by Taube (1953). In the Netherlands the literature is very sparse: mention can be made of the publications of Groeneveld (1947, 1954) and Wesseling (1963).

References

Anonymous (1955). 'The Need for a Faceted Classification as the Basis of all Methods of Information Retrieval.' *Int. Adv. Comm. Doc. Terminol. in Pure and Applied Sciences* (UNESCO 320/5515, Paris). Also in *Libr. Ass. Rec.* **57** (July), 262

Ariès, Ph. (1957). 'Index permanent des matières 1956.' *Fruits tropicaux et subtropicaux.* **20** (9bis) (Special issue) Paris; I.F.A.C.

Atherton, Pauline (ed.) (1965). *Classification Research. Proc. 2nd Int. Study Conf. Elsinore.* Copenhagen; Munksgaard

Batty, C. D. (1969). 'The Automatic Generation of Index Languages.' *J. Docum.* **25**, No. 2, 142

Baulkwill, W. J. and Posnett, N. W. (1969). *Provisional Thesaurus of Land Research Terms.* Tolworth, Surrey; Land Resource Division—Directorate of Overseas Surveys

Blagden, J. F. (1968). 'Thesaurus Compilation Methods. A Literature Review.' *ASLIB Proc.* **20**, No. 8, 345

Bohnert, L. M. (1955). 'Two Methods of Organizing Technical Information for Search.' *Am. Docum.* **6**, No. 3, 134

Borko, H. (1968). 'The Conceptional Foundation of Information Systems.' In: Montgomery, E. B. (ed.). *The Foundations of Access to Knowledge. A Symposium.* Syracuse; Syracuse Univ. Press

Bornstein, H. (1961). 'A Paradigm for a Retrieval Effectiveness Experiment.' *Am. Docum.* **12**, 254

Brisch, E. G. (1955). 'Subject Analysis in 81 Concepts.' *ASLIB Proc.* **7**, No. 3, 157

Broadfield, A. (1946). *The Philosophy of Classification.* p. 102. London; Chatterji, N. N. (1952). 'A Review of Literature on Colon Classification.' *Abgila* **2**, No. 6, 162

Cleverdon, C. W. (1962). *Report on the Testing and Analysis of an Investigation into the Comparative Efficiency of Indexing Systems.* College of Aeronautics, Cranfield, Bucks.

Cleverdon, C. W. (1967). 'The Efficiency of Index Languages.' In: *Communication in Science* (Ciba symposium). London; Churchill

Cleverdon, C. W. and Mills J. (1963). 'The Testing of Index Language Devices.' *ASLIB Proc.* **15**, No. 4, 106

Coates, E. J. (1960). *Subject Catalogues, Headings and Structure.* London; Library Association

Cuadra, C. A. and Katter, R. V. (1967). 'Opening the Black Box of "Relevance".' *J. Docum.* **23**, No. 4, 291

Doyle, L. B. (1962). 'Indexing and Abstracting by Association.' *Am. Docum.* **13**, No. 4, 378

Egan, M. E. (1953). 'Subject Headings in Specialized Fields.' In: Tauber, M. (ed.). *Subject Analysis of Library Materials.* pp. 83–99. New York; School of Library Science Columbia Univ.

Fairthorne, R. A. (1970). 'Content Analysis.' *A. R. Inf. Sci. Technol.* **4**, 73

Farradane, J. E. L. (1950). 'A Scientific Theory of Classification and Indexing and its Practical Applications.' *J. Docum.* **6**, No. 2, 83

Farradane, J. E. L. (1952). 'Further Considerations.' *J. Docum.* **8**, No. 2, 73

Foskett, D. J. (1961). 'Classification and Integrative Levels.' In: *Sayers Memorial Volume.* pp. 136–149. London

Foskett, D. J. (1962). 'Classification.' In: *Handbook of Special Librarianship and Information Work.* pp. 83–131. London; ASLIB

Foskett, D. J. (1963). *Classification and Indexing in the Social Sciences.* London; Butterworths

Gilchrist, A. (1971). *The Thesaurus in Retrieval.* London; ASLIB

Groeneveld, C. (1947). 'Problems of Classification.' *Revue Docum.* **14**, 99

Groeneveld, C. (1954). 'Over de grondslagen van de classificatietheorie.' *Bibliotheekleven* **39**, No. 2, 221

Grolier, E. de (1956). *Théorie et practique des classifications documentaires.* Paris; UFOD

Holmstrom, J. E. (1947). 'A Classification of Classifications.' *FID 17, Conference Rapports* **1**, 29

Jonker, F. (1959). 'The Descriptive Continuum: A Generalised Theory of Indexing.' In: *Proc. I.C.S.I.* Vol. 2, pp. 1291–1312. Washington

Kent, A. (1971). *Information Analysis and Retrieval.* New York; Becker and Hayes

Kyle, B. (1956). 'E. G. Brisch, Something New in Classification.' *Spec. Libr.* **47**, No. 3, 100

Lamontagne, L. E. (1953). 'Historical Background of Classification.' In: Tauber, M. (ed.). *Subject Analysis of Library Materials.* pp. 16–28. New York; School of Library Science Columbia Univ.

Lancaster, F. W. (1964). 'Mechanized Document Control: A Review of Some Recent Research.' *ASLIB Proc.* **16**, No. 4, 132

Lancaster, F. (1968). *Information Retrieval Systems.* New York; Wiley

Lorch, W. T. (1970). 'Internationale Richtlinien für Thesauri (Unisist).' *Nachr. Dokum.* **21**, No. 2, 72

Lustig, G. (1969). 'Ist die automatische Indexierung anwendbar?' *Nachr. Dokum.* **20**, No. 5, 190

Mandersloot, W. G. B., Douglas, E. M. B. and Spicer, N. (1970). 'Thesaurus Control and the Selection, Grouping and Cross Referencing of Terms for Inclusion in a Coordinate Index Word List.' *J. Am. Soc. Inf. Sci.* **21**, No. 1, 49

Mooers, C. N. (1956). 'Zatacoding and Developments in Information Retrieval.' *ASLIB Proc.* **8**, 3

Moss, R. (1967). 'Minimum Vocabularies in Information Indexing.' *J. Docum.* **23**, No. 3, 179

Needham, R. M. and Sparck Jones, K. (1964). 'Keywords and Clumps: Recent Work on Information Retrieval at the Cambridge Language Research Unit.' *J. Docum.* **20**, No. 1, 5

Palmer, B. I. (1953). 'Classification.' *Libr. Trends* **2**, No. 2, 236

Palmer, B. I. and Wells, A. J. (1951). *The Fundamentals of Library Classification.* London; Allen & Unwin

Proceedings (1957). *Proc. Int. Study Conf. Classification Inf. Retrieval.* London; ASLIB

Rahmann, M. and Samulowitz, H. (1968). 'Planung und Aufstellung eines Fachthesaurus.' *Nachr. Dokum.* **19**, No. 6, 222

Ranganathan, S. R. (1951). 'Colon Classification and its Approach to Documentation.' In: *Bibliographical Organization.* pp. 94–108. Chicago; University Library School

Rolling, L. (1965). 'The Role of Graphic Display of Concept Relationships in Indexing and Retrieval Vocabularies.' In: Atherton, P. (ed.). *Classification Research. Proc. 2nd Int. Study Conf. Elsinore.* pp. 295–325. Copenhagen; Munksgaard

Scheele, M. (1967). *Wissenschaftliche Dokumentation.* Schlitz/Hessen; Scheele

Schüller, J. A. (1964). 'De nieuwe thesaurus van het technisch documentatie-informatie-centrum voor de Krijgsmacht.' *Tijdschr. Effic. Docum.* **34**, No. 2, 84. Also published in *Proc. 26th Annual Meeting of the American Documentation Institute.* Oct. 1963

Shera, J. H. (1951). 'Classification as the Basis of Bibliographic Organization.' In: *Bibliographical Organization.* pp. 72–93. Chicago; University Library School

Soergel, D. (1969). *Klassifikationssysteme und Thesauri.* Frankfurt; Deutsche Gesellschaft für Dokumentation E.V.

Stevens, M. E. (1965). *Automatic Indexing. A State of the Art Report.* Washington; National Bureau of Standards (NBS Monograph 91)

Taube, M. (1951). 'Functional Approach to Bibliographic Organization: a Critique and a Proposal.' In: *Bibliographical Organization.* pp. 57–71. Chicago; University Library School

Taube, M. (1953). *Studies in Coordinate Indexing.* Bethesda, Md.; Documentation Inc.

Vickery, B. C. (1951). 'Some Comments on Mechanical Selection.' *Am. Docum.* **2**, No. 2, 102

Vickery, B. C. (1954). 'Books are for Use.' *Libri* **4**, No. 3, 265

Vickery, B. C. (1955). 'Developments in Subject Indexing.' *J. Docum.* **11**, No. 1, 1

Vickery, B. C. (1959). *Classification and Indexing in Science.* London; Butterworths

Vickery, B. C. (1960a). *Faceted Classification. A Guide to Construction and Use of Special Schemes.* London; ASLIB

Vickery, B. C. (1960b). 'Thesaurus, a New Word in Documentation.' *J. Docum.* **16**, No. 4, 181

Vickery, B. C. (1962). 'Classification for Documentation.' *ASLIB Proc.* **14**, No. 8, 243

Vickery, B. C. (1963). 'Vocabularies for Coordinate Systems.' *ASLIB Proc.* **15**, No. 6, 170

Wesseling, J. C. G. (1963). 'Classificatie als sleutel tot kennisverwerving.' The Hague; NIDER Publ. No. 40, 2nd series

THIRTEEN

TRANSLATION OF THE CONDENSATE INTO THE LANGUAGE OF THE RETRIEVAL SYSTEM
(Coding, notation)

Various reasons have prompted documentalists to employ signs or figures rather than words from everyday language. According to the situation, this is called notation, coding and so on. The treatment below is arranged according to the motive behind the translation of words 'in clear' into sign or number language. A review of current notational methods is given by A. C. Foskett (1969).

MOTIVE *A*: SHORTER EXPRESSION THAN LANGUAGE. SHORTHAND SIGNS

The need to shorten language by means of signs is much older than documentation. Sign languages of chemistry, mathematics, chess, music and shorthand all show this, but documentation has induced disciplines to seek a formalisation in this direction. An example of this is the S(ymbolic) S(horthand) S(ystem) by Selye for the field of medicine: this system is described by Ember and Padmanabhan (1962) and de Grolier (1962). The latter also mentions a similar system, the 'lography' method for biology by Tchakhotine. Gerr (1964) proposes a universal symbol language for science and technology. Bourne (1963) gives detailed accounts of notations derived from abbreviations (English).

MOTIVE *B*: FACILITATING SEARCHING STRATEGY BY NOTATION

In the construction of words in the normal language hierarchic linkages are not immediately apparent. However, notations can lead the researcher to the more general, the more specialised or the juxtaposed terms, which the words of language cannot always do. It is not a matter of course to go from 'spelt' to 'wheat' or even 'grain' but the UDC notations 633.113, 633.11 and 633.1 clearly show the way. Notations of faceted classification also belong here. The shorter the hierarchies (facets), the shorter the notation can be. Further reading in this field can be found in Coates (1957). Much shorter than the figures of UDC is the letter-type notation proposed by Dobrowolski (1965).

MOTIVE *C*: STANDARDISATION OF TERMS AND TERM LINKAGES FOR THE PURPOSE OF MECHANISATION

Melton (1958) has constructed a standardised machine language (for telegraphic abstracts) because colloquial speech has formed its individual words in such an unsystematic, irregular and haphazard fashion. The units, each of four letters, are constructed to a strict logic. She uses 214 'semantic factors' as elements and this artificial language approximates to the Chinese language, which also constructs long words from many short roots. Vickery (1959) has described this semantic coding and compared it with other notations.

MOTIVE *D*: NOTATIONS BY WHICH MATHEMATICAL PROCESSES CAN BE EFFECTED

The 'prime number coding' of Lamm (1961) belongs to this group, as do the notations from the group algebra of Boole (Taube, 1959). These characteristics are of special importance in putting machines to work.

MOTIVE *E*: POSSIBLE AUTOMATIC ARRANGEMENT

With individual words only automatic alphabetical arrangement is possible; with UDC, for example, automatic systematic arrangement is possible. In this case 'automatic' is used to cover clerical or machine activity.

Possible notations are those consisting of numbers, letters, signs and combinations of these.

The form of symbol is only of secondary importance and is generally regarded as a derived element, not playing any major role in the choice of retrieval system. Attention is directed to Vickery (1960) and D. J. Foskett (1963), who have made very considerable contributions to the study of notation. In the UDC system controversy much of the argument revolved around notation, i.e. around only one aspect. Thus *all too much importance was placed upon the numerical notation.*

In this field one difficulty of terminology still remains: 'coding' can mean *either* translation into a notation *or* translation into a machine-readable form—that is, preparation of input for a mechanised retrieval system. The first sense is that understood in this chapter; the second is dealt with under Programming in Chapter 18.

MOTIVE *F*

With notations all translation problems may be avoided. This is the main point against the thesaurus, which makes translation very difficult. Notation can be quite internationally understandable. This is also one of the main weapons of protagonists of UDC.

References

Bourne, C. P. (1963). *Methods of Information Handling*. Ch. 3. New York; Wiley

Coates, E. J. (1957). 'Notation in Classification.' *Proc. Int. Study. Conf. Classification Inf. Retrieval.* pp. 51–64. London; ASLIB

Dobrowolski, Z. (1965). 'Notational System with Short Symbols.' pp. 131–150. In: Atherton, P. (ed.). *Classification Research. Proc. 2nd Intern. Study. Conf. Elsinore.* Copenhagen; Munksgaard

Ember, G. and Padmanabhan, N. (1962). 'Symbolic Shorthand Systems for Physiology and Medicine.' *Methods Inf. Medicine* **1**, No. 4, 138

Foskett, A. C. (1969). *The Subject Approach to Information.* London

Foskett, D. J., *Classification and Indexing in the Social Sciences.* London; Butterworths

Gerr, S. (1964). 'A "scripta franca" for Science and Technology.' *Rev. Int. Docum.* **31**, 16

Grolier, E. de (1962). *A Study of General Categories Applicable to Classification and Coding in Documentation.* Paris; UNESCO

Lamm, E. (1961). 'Prime Number Coding.' *Am. Docum.* **12**, 172

Melton, J. (1958). 'Procedures for Preparation of Abstracts for Encoding.' In: Perry, J. W. and Kent, A. *Tools for Machine Literature Searching.* pp. 69–109. New York; Interscience

Taube, M. (1959). 'The Distinction Between the Logic of Computers and the Logic of Storage and Retrieval Devices.' In: *Studies in Coordinate Indexing.* Vol. 5, Ch. 2, pp. 15–29. Washington; Documentation Inc.

Vickery, B. C. (1960). *Faceted Classification. A Guide to Construction and Docum.* **10**, 234

Vickery, B. C. (1960). *Faceted Classification. A Guide to Construction and Use of Special Schemes.* London; ASLIB

FOURTEEN

SHORTCUTS IN CERTAIN PREPARATORY PROCESSES

There is a complication in the flow from document to retrieval system in that *all* the activities, *A*, *B* and *C*, shown in *Figure 20* (p. 90), do not always take place. Situations do arise where, in practice, activities can be omitted and therefore certain transformation processes are not necessary.

SHORTCUTS IN CONDENSATION (*A*/*B*)

From the mechanisation report of the Library of Congress (1963) it is clear that in the U.S.A. the future trend of document analysis will shift from methods *with* to methods *without* condensation, but this will only occur after many years. The whole document would then be brought unabridged into the system. There is no consensus of opinion as to the utility of normal language in document analysis. Farradane (1963) disapproves of normal language as a starting point. Yngve (1960), on the other hand, describes language as 'a beautiful instrument of precision' standing at our service and recommends integral text input in order that no loss of information may result. If the whole document is going to be brought into the system, this is only feasible if storage and retrieval are mechanised and the expenditure of time occurs in the output stage. As regards input, there are now typewriters which simultaneously translate the written text into binary code (tape typewriters); in the not-too-distant future machines might be able to read printed, typed or even written originals (character recognition machines). Document-alists are here in the same field as linguists and especially lexico-graphers of concordances (as, for example, Busa, 1958). Pietsch (1962, 1963) has stated that if in future the solution should be sought in total storage, rationalisation should start in the writing stage (that is, in the thought process) so that texts would be written in as compressed a form as possible and limited to essential results and descriptions. This brings us back to the argument of Farradane (1963), who would construct documentation, not on the written sentence structure of texts, but rather on human thought structures. Kent (1971) also allies himself to Pietsch and stresses order in *thinking* so that this would also be reflected in the documents.

Computer-memory storage of whole documents in information storage and retrieval systems is thus cheap, because document content representation is rendered superfluous. This low cost of input must, however, be gauged against the expense of output: a search of a whole-document store can only succeed if the question is formulated in the terms of the sought document, but, since the terms used in the document constitute an unknown in the search procedure, all conceivable words which might have been used in the language of the stored documents must be used in retrieval.

The fact is that living language—because it is living—survives all our attempts to standardise it for improving our manual or mechanical matching of enquiries with document texts. Moreover, the matching ability of a computer is nowhere near that of a person reading intelligently.

Another difficulty of retrieving information from a store of integral texts is that success depends on the language of the stored texts. If this language has the precision of chemical or mathematical formulae, scanning may succeed. But searching a text with *implications*, such as a literary essay, will give poor results, because implications depend on meanings *between* the lines (connotation) rather than meanings explicit in words *on* the lines (denotation). The more explicit the language used, the better retrieval will be; a rise in the number of implications means a decrease in retrieval effectiveness. In practice, this means that storage of complete texts should be more successful in natural sciences and technical disciplines than in the humanities and social sciences. These problems of exact sciences versus social sciences and humanities are also common to automatic indexing, extracting and abstracting (see also Chapter 11, p. 103).

Breton (1969) indicates that texts with high information value (laws, patents) are better candidates for storage of complete text than texts with lower information value such as popular science journals.

SHORTCUTS IN RELATION TO NOTATION (C)

The use of short cuts in the codification simply means that normal language is also the language of the retrieval system. (Examples: subject heading catalogues; the codeless scanning of Schenk, 1961; and the Unité systeme of Te Nuyl, 1958.)

References

Breton, J. M. (1969). 'Indexation par mots-clés on texte intégrale?' *Information et Question* **8**, 25

Busa, R. (1958). 'The Use of Punched Cards in Linguistic Analysis.' In: Casey, R. *et al.* (eds.). *Punched Cards.* New York; Reinhold

Farradane, J. E. L. (1963). 'Relational Indexing and Classification in the Light of Recent Experimental Work in Psychology.' *Inf. Storage Retrieval* **1**, 3

Kent, A. (1971). *Information Analysis and Retrieval.* New York; Becker and Hayes

Library of Congress (1963). *Automation and the Library of Congress.* Washington

Pietsch, E. (1962). 'Entwicklungstendenzen im Bereiche der Dokumentation und Information.' *Nachr. Dokum.* **13**, No. 4, 198

Pietsch, E. (1963). 'Die kuenftige Entwicklung der Dokumentation.' *Libri* **12**, No. 4, 287

Schenk, H. R. (1961). 'Der Einsatz von Lochkartenmaschinen und Computers in der Codeless scanning-method.' *Nachr. Dokum.* **12**, No. 1, 22

Te Nuyl, Th. W. (1958). 'L'Unité documentation system.' *Revue Docum.* **25**, No. 3, 65

Yngve, W. H. (1960). 'In Defence of English.' In: Kent, A. (ed.). *Information Retrieval and Machine Translation.* Ch. 40, pp. 935–940. New York; Interscience

FIFTEEN

THE TWO GROUPS OF RETRIEVAL SYSTEM TYPES

GENERAL CONSIDERATIONS

As already indicated, in research work usually the document is not identified by a single simple concept but rather by a number of concepts and thus a number of descriptors. This plurality gives rise to the question of permutation of these descriptors. There are two basic types of retrieval system:

1. Systems in which the descriptors are *combined in advance* by the constructor. These are systems with *pre*-coordination. To these belong the usual card indexes and book or periodical subject indexes. In these systems no machines (and therefore no binary code) are used for retrieval, as the collection is always arranged in a certain way (which entails sorting in) and does not need to be searched entirely (input costly, output comparatively easy).

2. Systems in which the descriptors are not combined in advance by the constructor. Here the descriptors are held in the system singly, not combined or 'coordinated'. These systems are generally named 'coordinate indexing systems' (Jaster, Murray and Taube, 1962) and are here referred to as systems with *post*-coordination. Many punch card systems and their mechanised forms are of this type. In some the entries are not arranged in order and thus have to be searched through; therefore, with the growth of the system, machines are used and translation into a binary code is required, but, on the other hand, there is no sorting in (input easy, output costly).

   In some 'term entry' systems, such as Uniterm, term cards require sorting in order, so as to be readily extracted for a search.

Two descriptors A and B can result in two possible descriptor permutations—$A:B$ and $B:A$ (the sign : is used here as in UDC). In the pre-coordinate system both of these permutations are entered. With post-coordination A and B are fed in singly; and if A and B (or B and A) are requested, the right target is reached in both cases. The compounding of descriptors, therefore, occurs not in the input but only with output. The order in which the terms of the query are put, i.e. the permutation, does not matter when one searches a system with post-coordination but does matter with pre-coordination: if only the permutation $A:B$ were entered there and the query is framed $B:A$, then the relevant document would not be found. In the case of two descriptors this does not represent a problem, since the two permutations can virtually always be written in. With three descriptors, however, the difference makes itself felt: the six possible permutations of A, B and C are not all written down in pre-coordinate systems. If a query were directed to a 'non-choice' permutation, the document would not be found. In the post-coordinate system permutation of the query does not matter, as the descriptors are stored singly, not pre-coordinately. In other words: all permutations are latently present in the post-coordinate system.

In the case of four descriptors (24 possible permutations) it is quite clear that only systems with post-coordination offer certainty, since in pre-coordinate systems, for reasons of 'efficiency', only a small selection of the 24 possible permutations can usually be entered into the system. Systems with pre-coordination depend on the choice between 24 possible permutations made by the constructor of the system. In the systems with post-coordination, therefore, a part of the input processing is transferred to output. From the cost–benefit point of view, this means that certain processes are carried out only when a query is raised. One can agree with Cleverdon (1962) that the difference between systems with pre- and post-coordination is important from the point of view of cost–benefit but is not a difference in principle.

The following conclusion can be reached: pre-coordination is fully satisfactory with one or two descriptors per publication; partially with three; with four or more, systems with post-coordination must be chosen.

The choice between the two systems would therefore appear to be very simple. There are, however, other factors at work:

(a) Free permutation of all descriptors is not always requested (see p. 134).
(b) There are factors reducing the number of descriptors per document (see p. 138).

(c) There are factors reducing required numbers of permutations (see p. 139).

These three factors can lead to the result that, because of the number of descriptors, a post-coordinate system would appear to be the first choice but, because of further factors, a pre-coordinate system is the actual decision.

These three factors are now considered.

THE QUESTION OF FREE PERMUTATIONS

The previous conclusion was based on the assumption that free permutation is always necessary. In two cases, free permutation is not necessary, i.e. when it is known that:

(a) The searchers seek entry by means of one descriptor permutation in one set grid of reference—for example, Plant–Disease–Country. There is already a built-in hierarchy in the query which obviates permutation; it also renders it unnecessary to transfer from the pre-coordinate to the post-coordinate system (see below).

(b) The searcher is prepared to plough through a whole group of descriptor combinations in the hope of finding *one* that is relevant. Here it is a case of saving permutations at the cost of search-time (with rotation index, chain index, KWIC and subject heading plus sentence system). In this respect, the academic researcher enjoys a certain freedom in browsing and generally puts up gladly with the longer search-time.

With regard to (a): if it is laid down in a system that a certain document is dealing with potatoes as pig fodder, this can be represented in UDC as 636.4.086.491. It is then retrievable under Pigs–Fodder–Potatoes but not under Potatoes or Fodder as first descriptor or under Potatoes as second descriptor. The document is only retrievable if a certain hierarchy of the criteria of division is observed—namely, that which is built into the classification system. If the searcher conforms to the grid, he finds the document. Free permutation is not required in such cases. It may be noted in passing that were these three possibilities to be sought by the UDC system the three codings could be given as

$$636.4 : 591.13 : 633.491$$

and all six permutations of these three descriptors then written in.

This means that the numbers become shorter and the hierarchic classification, which is built into the UDC, is less used; but, on the other hand, the searcher is less constrained.

With classic classifications, such as UDC, there are two modes of use:

1. The longer the numbers, the more hierarchy is built in and the less the colon sign : is used.
2. The shorter the numbers, the less hierarchy is built in and the more the colon sign : is used.

In the first case the classification aspect is fully exploited, in the second case less, which supports the tendency towards pure use of the decimal notation.

Classifications mean diminution of the need for permutation. The less classification, the more need for permutation.

There are also situations which call for a hybrid system, where free permutation is needed in one area and classification in another. In other words, it can happen that in one area monohierarchic queries are to be expected and in the second area of the same system queries without built-in hierarchy. In such cases the first zone can use a pre-coordinate, the second a post-coordinate system (de Weger, 1959).

With regard to (b), the saving of permutations at the cost of search-time: Five examples of these systems follow: they are all systems which require lengthy search-time, offer greater browsing possibilities and do not necessitate any free permutations. Vickery (1959) suggests that it is reasonable to assume that during a search 'your eye falls upon' a recognised descriptor or combination of descriptors appearing closely together.

(1) Rotation Index (cyclic permutation)

In the rotation index each descriptor stands once only in first place. The sequence of descriptors remains (cyclic) always the same. Thus, instead of 24 permutations of A, B, C and D, there are only four:

$$D:C:B:A \quad B:A:D:C$$
$$A:D:C:B \quad C:B:A:D$$

Campbell (1963) is of the opinion that complete permutations are only worth while in a few cases and that the rotating of four descriptors mostly suffices. 'To enter more permutations than this, except for very important material, is of doubtful economic value.'

(2) Chain Index

A chain index is a system which represents the publication at one place only and with a fixed descriptor sequence only (for example, $A:B:C:D$) and references lead back to this entry:

$$
\left.
\begin{array}{l}
D:C:B:A \\
C:B:A \\
B:A \\
A
\end{array}
\right\} \quad \text{see } A:B:C:D
$$

An example of a chain index is reviewed by Dowell and Marshall (1962). Detailed descriptions of the chain index are to be found in Vickery (1959), Coates (1960) and Batty (1970). Cleverdon (1962), D. J. Foskett (1962) and Bakewell (1968) are critical of this system. Lancaster (1968) indicates that systems with a preferred order and chain indexing go hand in hand, as in preferred order the most general term normally precedes the less general and in chain indexing it is the other way round.

(3) KWIC

In K(ey) W(ord) I(n) C(ontext), as used in the indexes to *Chemical Titles*, the same sequence of words is always followed but it is not the first word that is scanned but words in a central column. If, for example, the sequence $D:C:B:A$ were taken as the words in the title, the various points of access in KWIC would be expressed as follows:

		Scanning column	
$D: C: B:$		A	
	$D: C:$	B	$:A$
	$D:$	C	$:B:A$
		D	$:C:B:A$

Alphabetical order is maintained in the scanning column. However, the researcher who has found his first entry word must now look through all combinations including this word, which under certain circumstances can be a considerable task.

A survey of the use of the KWIC index is given by Schneider (1963). A description of the setting up of a KWIC index and a critical review of this method is given by Black (1962).

D. J. Foskett (1962) deals with methods (1), (2) and (3); he considers complete permutation to be uneconomic. In his comparative studies he arrives at a special recommendation of (1) (rotation index), classification symbols being used instead of words.

(4) Alphabetico-specific System

The alphabetico-specific system is used in the indexes to *Chemical Abstracts*. The first descriptor is followed by a short sentence and the researcher must look through all such sentences. Search-time is longer, the possibility of browsing is greater and the retrieval is sharper.

(5) SLIC Index (Selective Listing In Combination)

Sharp (1964, 1965) describes a system in which combinations of terms are derived from the set of terms assigned by the indexer, in alphabetical order: only those groups are used (2^{n-1}, where n is the number of terms) which do not form the beginnings of longer groups, and each term appears next to every other term.

(6) The Helion System

Helion (1970) uses a file with descriptors in which every descriptor is accompanied by references to the descriptor combinations in which this descriptor is used in the system. This first file and these references make abundant permutations superfluous but imply the use of one permutation only for each combination. By this doubling of the search file the problem of the free permutations is solved in simple manner; to achieve the same effect with systems with post-coordination, a special device must be used.

For further discussion of the question of free permutations see the book by A. C. Foskett (1969).

CAUSES OF THE DIMINUTION OF THE NUMBER OF DESCRIPTORS PER DOCUMENT

If a document is to have eight descriptors in all and the post-coordinate system has been chosen, this simply means that it is assumed that the document concerned contains information about *all* possible permutations of these eight descriptors. In practice, however, on closer examination a document divides into units in which the linkages between two, three or four descriptors only are dealt with. If, despite this, the document is regarded willy-nilly as *one* unit and entered into the system with eight descriptors, then numerous superfluous correlations would be introduced which, in reality, are not present in the document at all. In *Figure 28* it can be seen how great the difference can be between the theoretical number of linkages using all eight descriptors and the real number of linkages actually treated in the document. This 'redundancy' can be countered by dividing the document into smaller units, in each of which the descriptors can be used in real linkage. The descriptors related to the same smaller unit are marked with the same 'link'.

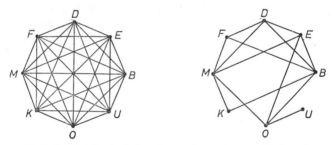

Figure 28. (left) The theoretical and complete scheme of linkage between eight descriptors. (right) A possible scheme of linkage between the same eight descriptors in practice

The introduction of these smaller units (microthemes; see also Chapter 10)—that is to say, sections from the document—diminishes the number of descriptors per unit and, hence, the necessity for using a post-coordinate system.

In such cases it may well be possible that only a part (be it a paragraph or even only a sentence) is extracted, leaving the surrounding context undocumented as irrelevant. Even in the so-called 'collectanea' methods (Hines, 1961) only a small part (a sentence, for example) is extracted. Meyer (1961) suggests documenting with partial concept chains, 'building blocks'.

CAUSES OF THE DIMINUTION OF THE NECESSARY NUMBER OF PERMUTATIONS

If a system is still limited and small, a less profound construction suffices; not all permutations have to be written out. Thus, with four descriptors there are 24 permutations but if access with two descriptors is sufficient, then in the same case only 12 permutations have to be written out.

INCREASE IN DESCRIPTORS

With the semantic factoring method (Perry, 1956) there is an increase in the number of descriptors. This method divides a descriptor into descriptors of more elementary terms—thus, thermometer into: instrument, measurement, temperature.

The great advantage of this method is that the list of descriptors becomes smaller, or remains relatively short: mostly so short that it can be perused. Against this, the number of descriptors per document increases, so that a post-coordinate system becomes all the more necessary.

THE APPLICATION OF A POST-COORDINATE SYSTEM

When does a situation arise with more than three descriptors and free permutation without any diminution factors? In the case of a central object or problem, either (1) with many independent variables of property or (2) with factors mutually influencing the central problem and one another.

Examples of (1) are: a system of chemical compounds and their properties; an index of patients with various data for a hospital administration; a population card catalogue with various demographic data. An example of (2) is: a system for tomato diseases with the factors which influence this state.

It is clear that these systems can only have queries posed to them with the same central objects or problems. Such systems are mostly prepared for this purpose. This situation often arises in research where documentation is being carried out to solve one exactly defined problem (mission-oriented). In this data documentation the documents are, so to speak, *prefabricated* for documentation and generally the document consists only of one card with data.

In all systems used in libraries such a unified query approach is missing. Therefore, in this case there can be no question of a *single* central object or problem. The queries which arise in a library must, of necessity, be confined to the level of the central problem *itself*. The number of descriptors of this central problem is usually smaller than those of the independent properties or factors already mentioned.

In Chapter 10 under 'Queries' the five practical situations are named: (a) public library, (b) university library, (c) institute library, (d) a problem for a research team, (e) a collection of data from an investigation. In (a) and (b) the unified query approach is missing; in (c) the situation is much the same as in (a) and (b); only in (d) and (e) is the situation such that the type of the normalised queries might justify the application of a post-coordinate system.

CONCLUSIONS

The choice between a pre- and post-coordinate system is not simple, as we have seen. Post-coordinate systems tempt by their very modernity; they count as major progress and yet have caused much disappointment. It has been shown that there are cases where free permutation is not necessary; that analysis into units of thought, reducing the descriptors per unit, renders post-coordinate systems necessary in fewer cases; that, finally, the smallness of the system itself does not require a post-coordinate approach.

Confusion is, however, magnified by the fact that post-coordinate systems are often used for reasons *other* than those described above (see Chapter 18 under 'Automation').

It is therefore clear that early considerations concerning document analysis will influence the choice between pre- and post-coordinate systems. Such considerations include size of unit selected for documentation or measure of selected condensation and thesaurus (amount of built-in hierarchic structure or the attempt to keep the thesaurus as short as possible). In addition, there are other factors (for example, size of the system and amount of search-time available to the researcher) which determine the choice of system.

References

Bakewell, K. G. B. (1968). *Classification for Information Retrieval*. London; Bingley

Batty, C. D. (1970). 'Chain Indexing.' Encycl. Libr. Inform. Sci. **4**, 423

Black, J. B. (1962). 'The Keyword.' *ASLIB Proc.* **14**, No. 10, 313

Campbell, D. J. (1963). 'Making Your Own Indexing System in Science and Technology.' *ASLIB Proc.* **15**, No. 10, 282

Cleverdon, C. W. (1962). *Report on the Testing and Analysis of an Investigation into the Comparative Efficiency of Indexing Systems.* College of Aeronautics, Cranfield, Bucks.

Coates, E. J. (1960). *Subject Catalogues. Headings and Structure.* London; Library Assoc.

Dowell, N. G. and Marshall, J. W. (1962). 'Experience with Computer-Produced Indexes.' *ASLIB Proc.* **14**, No. 10, 323

Foskett, A. C. (1969). *The Subject Approach.* London; Bingley

Foskett, D. J. (1962). 'Two Notes on Indexing Techniques.' *J. Docum.* **18**, 188

Helion, J. (1970). 'Le classement par mots clefs et l'archivage sous forme de microcopies au service documentation de l'Institut technique de l'élévage bovin.' In: *IAALD. IVth World Congress.* pp. 247–249. Paris; Institut National de la Recherche Agronomique

Hines, Th. C. (1961). 'The Collectanea as a Bibliographical Tool.' Ph.D. thesis. Graduate School of Library Science, Rutgers The State University, New Brunswick

Jaster, J. J., Murray, B. R. and Taube, M. (1962). *The State of the Art of Coordinate Indexing.* Washington; Documentation Inc.

Lancaster, F. W. (1968). *Information Retrieval Systems; Characteristics, Testing and Evaluation.* New York; Wiley

Meyer, E. (1961). *Grundfragen der Patent dokumentation.* Düsseldorf; V.D.I. (Especially Section 318, 'Documentation of Patents at the Dividing of the Ways.')

Perry, J. W. (1956). *Machine Literature Searching.* New York

Schneider, K. (1963). 'Fünf Jahre KWIC—Indexing nach H. P. Luhn.' *Nachr. Dokum.* **14**, No. 4, 200

Sharp, J. R. (1964). *The Slic Index.* ASLIB Annual Conf. pp. 2–11, Exeter

Sharp, J. R. (1965). *Some Fundamentals of Information Retrieval.* pp. 205–213. London

Vickery, B. C. (1959). *Classification and Indexing in Science.* London; Butterworths

Weger, A. de (1959). 'Bedrijfsdocumentatie.' *Verandwoord toegankelijk Maken van Literatuur.* The Hague; NIDER Publ. No. 24, 2nd series

SIXTEEN

THE FOUR TYPES OF RETRIEVAL SYSTEM

INTRODUCTION

From Chapter 10 to this point, it has been suggested that the choice
of a particular retrieval system has been based on the choice of
types of descriptor. Now we consider a further differentiation, in
which the information carrier I and the document address A play
their part.

It has been shown that the basic type (I) of an element is
$I:D+A$. There are only two derived types, II and III (by means of
condensation), which are also of importance, namely $I:xD+A$ (II)
and $I:D+yA$ (III). The derivation of II from I can be represented
by a document address being combined with several (x) descriptors
(D', D'', D^x) in Type I; there are then x information carriers
$I:D+A$, $I:D'+A$, $I:D''+A$, $I:D^x+A$. If all these combinations
are brought together on a *single* information carrier, Type II results,
$I:xD+A$.

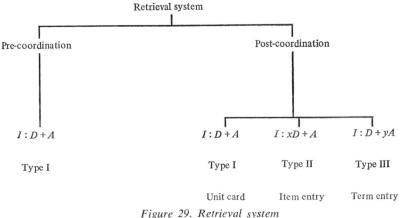

Figure 29. Retrieval system
142

The derivation of III from I can be represented by a descriptor being combined with several (y) document addresses: in Type I there are then y information carriers: $I:D+A, I:D+A', I:D+A''$ ... $I:D+A^y$. If all these combinations are brought together on a *single* information carrier, Type III results, $I:D+yA$.

THE ONLY TYPE IN SYSTEMS WITH PRE-COORDINATION

Only Type I is possible in systems with pre-coordination: if we take a document address that has to be linked with three descriptors (D, D', D'') the corresponding cards in a pre-coordination system show the following structure:

$$I:[D \ :D' \ :D'']+A$$
$$I:[D \ :D'':D' \ ]+A$$
$$I:[D' \ :D \ :D'']+A$$
$$I:[D' \ :D'':D \ ]+A$$
$$I:[D'':D' \ :D \ ]+A$$
$$I:[D'':D \ :D' \ ]+A$$

All these six permutations are raised on six separate cards which are themselves of Type I. Every permutation of descriptors plays the role of one descriptor in this context.

THE THREE TYPES IN SYSTEMS WITH POST-COORDINATION

The following nomenclature will be used to describe the three types:

Type I	$I:D+A$	Unit card
Type II	$I:xD+A$	Item entry
Type III	$I:D+yA$	Term entry

We first consider Types II and III, item and term entry. The difference between the two systems can best be grasped by referring to the classic card systems. The riders in such card systems are universally known: they represent specific properties of that which is shown on the card, whereby the colour and position of the rider can also have some meaning.

Let it be imagined that the properties would be registered by holes *in* the cards rather than riders on the cards. It can further be visualised that transference is not by means of holes to cards but to transparent and opaque (black and white) parts of film material. Item entries of magnetic or electronic type can easily be visualised. This whole group is called the item entry system. Each document has a card (or a carrier of other material) as a unit with a distinct item entry.

If the term entry system were to be derived from the classic card catalogue system, it would have to be imagined that such a card catalogue was concentrated on the (subject) guidecard. Thus all titles (naturally abbreviated, to a simple numbering, for example) of the documents, whose cards follow the guidecard, would be brought together on the guidecard itself. The whole system consists only of (subject) guidecards with numbers (the so-called 'inverted file'). The cards are therefore subject heading cards with accession numbers. Publications on combinations of topics are searched for by extracting the relevant subject cards from the system and seeking on them the same numbers. In order to avoid this time-consuming comparison, other methods are being employed, which will be discussed below. *Figure 30(a)* shows the item entry system, in which a card is raised for each publication and in which, for each access to the subject, a hole is punched. In searching, a certain number of field places are set and sorted out (by needle or mechanically) showing which cards correspond to the query item entry. *Figure 30(b)* shows the term entry system: a punch card is raised

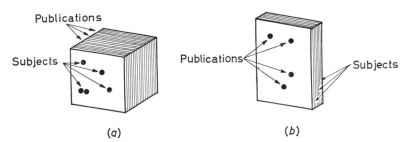

(a)   (b)

Figure 30. Schematic comparison between item (a) and term entry (b) systems. (After Taube, 1955)

and a punching is made for each publication. In searching, the relevant subject punch cards are superimposed and held against the light. Where the light shines through, a publication number is recalled which is fully relevant to all the queries stated. A good exposition of these differences also appears in Jolley (1960).

The essential difference between item and term entry systems will become apparent during their construction and growth. New documents do not present a problem for the item entry system: for each new document a new card is raised. However, the accommodation of new terms can be difficult, because of the restricted card surface. On the other hand, in the term entry system the accommodation of new *documents* is restricted by the card surface.

Term entry systems enable one to follow the dynamics of scientific research step by step, not only because of the addition of new cards at any stage in the building up of the system but also because of the ease with which cards can be combined in order to reduce the original specificity or a single card can be divided into different cards to increase the specificity. It is just this great flexibility that is so important for research workers.

The disadvantages can be mitigated in both cases to some extent by a different use of the sole limiting factor, namely the surface area of the carrier; and where greater quantities of documents are to be processed, then item entry is the system of choice.

A second, not so essential but for practical purposes even more important, difference between item and term entry systems lies in the fact that, in item entry, codification of terms becomes a necessity—for example, into the punch code of the punch cards; a processing step (which is superfluous, for obvious reasons, in the term entry system) must be interpolated. This processing step can lead to loss of definition and means, in addition, extra work in the data input stage. This demands certain processes which enable the step to be made from notation to punch pattern (for the following step—punch pattern–punch card, i.e. the actual input—attention should be paid to the cost–benefit management point of view, since this is purely a techno-economic matter).

Fugmann (1962) points out that this disadvantage of item entry systems can actually be turned into an advantage, guiding the searcher from the notation towards more comprehensive, or more specialised, terms being built into the system. This building-in process can also be achieved in the term entry system by means of double punching but is much more cumbersome. Fugmann, therefore, suggests that item entry systems should be preferred but that from them 'stills' should be taken—that is to say, lists of documents per term. On the other hand, term entry systems take up more search-time, since only numbers are retrieved and not subject indicators, as on the cards of item entry systems.

In addition to the item and term entry systems there is Type I, the unit card system. Here the advantages of the other two systems are combined: for each document and each descriptor a card is raised. In this way, in each direction, the freedom of interpolation of documents and terms is preserved. However, the number of cards becomes so great that automation of retrieval is the only solution.

An outline comparison of the three systems is presented in *Table 9*.

Table 9

System	*Increase of documents*	*Increase of terms*	*Processing*	*Automation*
Item entry	Free	Restricted	Necessary	Possible
Term entry	Restricted	Free	Not necessary	Possible
Unit card	Free	Free	Not necessary	Necessary

CONCLUSION

Item entry systems are particularly suited to large, rapidly expanding collections, in fields enjoying a certain stability. Term entry systems prove their worth in smaller collections with restricted intake, in fields with dynamic development. Unit cards come in for consideration for major collections with massive intake in documents as well as new terms.

It can be seen that, where the card surface area or, more generally, the area of the information carrier plays an important role, as discussed above, many attempts have been made to reduce this restriction (see also Chapter 17).

References

Fugmann, R. (1962). 'Ordnung und oberstes Gebot in der Dokumentation.' *Nachr. Dokum.* **13**, No. 3, 120

Jolley, J. L. (1960). 'Data Handling by Card Manipulation.' *J. Docum.* **16**, No. 3, 132

Taube, M. (1955). 'Storage and Retrieval of Information by Means of Association of Ideas.' *Am. Docum.* **6**, No. 1, 1

SEVENTEEN

FURTHER DIFFERENTIATION OF SYSTEMS WITH POST-COORDINATION

There are many books describing in detail the various systems with post-coordination as their mode of function and the types commercially available. There is no need, therefore, to repeat these descriptions. Reference can be made to the works of Casey *et al.* (1958), Scheele (1959), Becker and Hayes (1963), Kent (1971), Bourne (1963), Williams (1966) and Vickery (1961, 1970), to name but a few of the leaders in this field. Important sources are: *International Conference on Scientific Information (ICSI) Proceedings*, Washington (1958) and the *ADIA Proceedings* (1961). The great interest in documentation taken in the U.S.A. even by top administrators is clearly seen in *Report* (1960) to U.S. Senate. The pioneers can well be said to be the group which, in the Center for Documentation Research, Cleveland (Ohio), for many years undertook investigations in this field, when documentation, let alone scientific research on documentation, was scarcely taken seriously. The names Perry, Kent and Shera should be mentioned in this connection (see Shera, Kent and Perry, 1956, 1957; Perry and Kent, 1957; and Perry, Kent and Berry, 1956). In the series *Advances in Documentation and Library Science*, reflections of this immense activity may be found. A further breakdown of the types of retrieval system listed in *Figure 29* is given on p. 142 *without* all the possible examples and illustration of variants.

Figure 31 outlines the further differentiation of systems with post-coordination.

TYPE I

The unit card system offers additional possibilities according to experiments in the U.S.A. reported by Whaley (1958, 1961) and

Peakes (1957). It is true that costly machine retrieval is inevitable in this system but apparently it is the most flexible type. Magnacard (Nelson, 1958) is a further example of the unit card. As Whaley (1961) shows, only a small space is required for coding on the unit card and this allows various data to be added to the remaining space.

TYPE IIA

The scan column index is a system in which, as the name implies, the documents are represented on a list with the addition of item columns variously indicated (by signs, letters or numbers). Searching consists of vertical scanning and the choice of a certain combination of signs, within a certain combination of columns (described by O'Connor, 1962). In principle, it is the same as searching a card

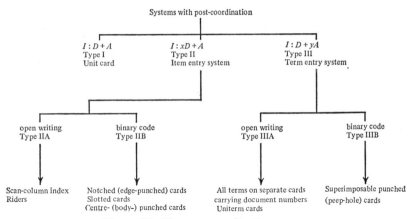

Figure 31. The post-coordination systems

catalogue, with riders of various colours along the tops of the cards at various points, for a combination of riders of a certain colour at certain points. This kind of scanning can be done much more quickly by machine than manually.

TYPE IIB

Binary codes can be accommodated by this type, as well as by Type IIIB. Previously this presented a major obstacle, since the translation of open text into binary code, e.g. punch card language, was so difficult, being well outside the scope of normal librarianship and documentation. This situation has changed radically now that typewriters are commercially available which can produce open and binary code text simultaneously. Another method which circumvents manual punching is the mark sensing card. On this card places for punching are indicated by a special pencil. Punching is then done automatically by a scanning machine. Here the work process from noting of punch position to punching itself is fully automated. In this context, character reading devices may eventually be used. Jans (1956) reports that punching charges for the same amount of cards costs 160 DM for machine punching by means of IBM, 32·5 DM for manual punching, but only 5 DM for mark sensing cards. The Flow Research Institute at Goettingen uses formulae employing the same system.

As to manual and machine systems in this group, the following should be noted: there are two manual systems, namely notched (edge punched) and slotted cards. The difference between them lies mainly in the mode of retrieval (Elsner, 1961). The advantage of the slotted cards is the possibility of more rapid retrieval, whereas the notched cards must on occasion be sorted on two, three or even four edges. Slotted cards are also less vulnerable than edge punched ones with their notched edges. On the other hand, notched cards generally provide more space for the text and no sorting apparatus is required. Against this, slotted cards require a shake-out arrangement or double needle sorter. In a needle sort with, for example, 150 cards each run, is theoretically possible to sort 150 000 cards manually—but it takes about a thousand times longer than a mechanical sort. For other comparative studies of these two systems see Elsner (1963).

(A) Notched (Edge-punched) Cards

There are different types of notched cards, with varying numbers of rows and edge punchings. The low capacity of the cards is often increased by complicated coding (for discussion of the triangular system and similar measures, see Chapter 18).

(B) Slotted Cards

Two different types of slotted cards are distinguishable. In the first type there are always two interlinked fields sorted by two corresponding needles which separate the cards into two packs (chosen and not chosen). The sorting needles are then removed and another pass is made with new needles on both sides at the point where only one of the two packs is raised. In the second type a number of cards are fixed by means of a needle and the others are either shaken out or lifted out. It is essential that the cards slide lightly and easily over one another; all must fall into place (no false drops). There are special shakeout boxes for this type of retrieval. The techniques of slotted cards are clearly described by Herrlich (1963).

(C) Mechanically Sortable Centre- (body-) punched Cards and Other Types

The mechanical systems of this group vary according to the form of the information carrier—card, paper, film, magnetic tape. Examples are: punch cards (card); punch tape (paper); the Filmorex of Dr. Samain, for example, and its variations, Minicard and microfilm retrieval of ERA (Engineering Research Association) (film). Examples of magnetic tape are to be found in the sector of the larger 'memories' in computer complexes.

The use of the usual 80 column punch card with mechanical sorting is often encountered, as in many firms or institutions there is a punch card department for accounts or statistics which, not being fully loaded, can be employed for literature searches. (The same can also be said of computers.) If the costs of such a department were to fall entirely on literature searching, they would be excessive for the most part. The Commonwealth Bureau of Plant Breeding, in Cambridge, provided the example of an economical mechanisation with punch cards by means of a small sorting machine specially purchased for this purpose. This institute has pioneered work with these punch cards since 1949 (Richens, 1958).

It is interesting to note the numerous reasons why (according to Ashthorpe, 1952) the English Atomic Energy Research Establishment gave up, after two years, its punch card system with mechanical retrieval:

(a) Too slow; too many cards had to be run through the machine to answer a single query.

(b) Too little room on the cards for the abstracts.

(c) In order to avoid searches on too narrow terms, generic terms were chosen; consequently the final sort had to be done manually after all.

(d) Too many stoppages because of worn cards. In one year 20 per cent of the cards had to be replaced.
(e) Original estimates of expected number of queries proved too high, and machines were idle.
(f) Impossibility of simultaneous query search.
(g) Impossibility of retrospective search or earlier codings.

This series of motives is very informative and indicates the evaluation which should necessarily precede the choice of a system. As already mentioned at the beginning of this chapter, there is an immense amount of literature on Type IIB, to which reference can be made in this context.

TYPE IIIA

The Type IIIA system is also known as Uniterm. A difficulty is the fact that the name 'Uniterm' is used in the literature for two quite different aspects of the system thus named by Taube (1953/64).

(a) 'Uniterm' stands for the special kind of descriptors (using only word roots in order to keep the number of descriptors employed as low as possible) proposed by Taube (see Chapter 11).
(b) 'Uniterm' also refers, however, to the way in which the cards are used: division into ten columns to accommodate accession numbers of the publications arranged by final number. This facilitates the comparison of two cards. Compared with superimposable punch cards, the comparison time is considerably longer but the card is more economically used, as only room for actual entries is required.

TYPE IIIB

Since the superimposable punch card is a truly flexible and fast working aid, whose potentialities for more modest documentation in restricted fields (by, for example, a research worker or a team or syndicate) have been far too often overlooked up to now, a few further points will be discussed.

Although the first patent dates from Taylor (1915), who used a primitive superimposable card device for the identification of birds by combination of characteristic properties, it was some time before both science and research recognised the great advantages of this system. Further details of the history of these cards are given by Kistermann (1957/58), Wildhack and Stern (1957) and de Grolier (1962). Particulars, in varying degrees, of the superimposable punch card system are given by Manual (1955), Bartels (1961), Boyd *et al.* (1963), Jolley (1959), Loosjes (1964) and Hermann and Löschner (1961).

The capacity of the individual card is restricted by its area, size of aperture and layout. Triangular layout means a saving of 16 per cent of surface area compared with square layout. Apertures smaller than 2 mm require precision apparatus for punching and recognition. The Delta card (*Figure 32*), with its 10 000 positions, has apertures of 2 mm and triangular layout and is thus the superimposable card with the smallest area accepting the greatest numbers of punchings, *without* requiring precision aids for punching and read-back (Loosjes, 1957). *With* precision instrumentation, cards having up to 40 000 positions can be used (Jonker, 1959). The well-known 80 column card (960 positions) can also be used for this purpose (Westendorp, 1956).

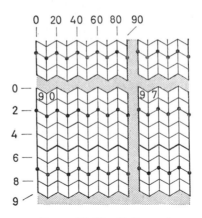

Figure 32. The Delta card

Cards of this type offer many interesting possibilities:

(a) The number of cards used can be reduced by superimposed coding. For example, 25 terms can be covered by 10 cards used in pairs (i.e. any one of the first 5 with any one of the second 5 gives 25 combinations). For a practical application see Schüller (1963) and Schüller and Koekkoek (1962). Müller (1970) discusses these possibilities at length and gives references.

(b) Possibilities of negative retrieval (*Figure 33*). If T is yellow translucent paper, it can be established which holes are covered by

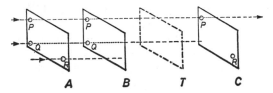

Figure 33. Negative retrieval

B and which by *C*. The holes must be viewed in this case under light from above and not from behind the cards.

(c) A group sorting is made possible by raising group cards, which are always punched simultaneously with the single cards of the group. There is also the possibility of a combination with edge-punched sorting for such cases (Kistermann and Úhlein, 1957).

(d) New term cards can be integrated at any time. Later it is possible to establish by the pattern of punches when (that is to say, at what accession number) the new term has been introduced.

(e) In the event of there being many series of different colours prepared at various times, it can be advisable not to arrange the cards by colour (chronologically) but according to terms, which considerably increases search speed (Westendorp, 1959).

(f) Volume number and page numbers can be used as coordinates on the card, the abbreviated periodical title being added in each case at the punch hole (Heinze, 1957).

(g) Experiments in combining superimposable punch cards and microfilm, 'Microcite', are discussed by Wildhack and Stern (1957). This microfilm contains bibliographic details which correspond to the publications' accession numbers indicated by the holes. By direct enlargement immediate bibliographic details are obtained in open text. Plankeel (1960) reaches the same result by means of a punched strip combined with filmstrips.

(h) Braband (1957) extends the use of superimposable punch cards by means of a basic card with colours (red, black) on the aperture positions. These colours give particulars of personnel contained in the card collection.

(i) Jolley (1963) describes a method of reference in the cards to associated descriptors by means of a reserved space where associated descriptors can be punched in by numbers (extended field for associated features).

(j) Searching and replacing cards can be simplified by means of a system with adjustable diagonal fields, as is usual in book-keeping (Engelhardt, 1958).

(k) The form of hole can be differentiated and the capacity of the card thereby increased (de Weger, 1959).

(l) A second system can be superimposed, in which each number does not refer to a publication but to a definite block of superimposable punch cards. In this manner this combined system acquires a squared capacity compared with the originally used superimposable punch cards (Ashworth, 1962; Shedder, in Boyd et al., 1963).

(m) Quantitative evaluations when the over-all pattern of the card is compared with the hole patterns of other superimposable punch cards.

(n) Fugmann (1962) discusses the advantages of a combination of superimposable cards with normal files which are not inverted.

Mechanised forms of the superimposable punch card system are commercially available under the names of Special Index Analyser and Ramac. Costello (1962) has discussed the advantages of automation of term entry systems. Hauser and Herrlich (1969) have experimented with systems of superimposable cards with large capacity and with the help of mechanical punching.

From the viewpoint of organisation, superimposable systems are especially practical with subscription to card services. Non-inverted systems oblige the subscriber periodically to do filing work. The new cards, arranged according to number and accompanied by a key on superimposable cards, do not need any further filing work. Both private and library subscribers, not geared to filing work on subscriptions, will welcome this method. It is practised by the documentation service in the agricultural field at Stuttgart Hohenheim (Germany). The superimposable cards that are distributed are printed on plastics, which is more economic than distribution of cards with punched holes.

Burkett (1968) in his review of applications of superimposable cards in the U.K. refers to 'increased acceptance'.

References

ADIA Proceedings (1961). *Proc. ADIA Conf. 1959, Frankfurt.* Ed. by. E. H. Pietsch. Suppl. *Nachr. Dokum,* No. 8

Advances (1956). *Advances in Documentation and Library Science.* Vol. 1. New York; Interscience

Ashthorpe, H. D. (1952). 'The Punched Card Indexing Experiment at the Library of the A.E.R.E. Harwell.' *ASLIB Proc.* 4, No. 2, 107

Ashworth, W. (1962). 'A Review of Mechanical Aids in Library Work.' In: *Handbook of Special Librarianship.* pp. 401–435. London; ASLIB

Bartels, W. (1961). 'Die Sichtlochkartei und ihre Folgeeinrichingen.' *Nachr. Dokum.* **12**, No. 2, 77

Becker, J. and Hayes, R. M. (1963). *Information Storage and Retrieval*; *Tools, Elements, Theories.* New York; Wiley

Bourne, C. P. (1963). *Methods of Information Handling.* New York; Wiley

Boyd, G. M., Childs, S. B., Johnson, A. and Shedder, D. A. (1963). 'Practical Applications of Feature Card Systems.' *ASLIB Proc.* **15**, No. 6, 178

Braband, C. (1957). 'Einige Vorschläge für die Ausgestaltung der Sichtlochkartenverfahren.' *Nachr. Dokum.* **8**, No. 1, 42

Burkett, J. (ed.) (1968). *Trends in Special Librarianship.* London; Bingley

Casey, R. S., Perry, J. W., Berry, M. M. and Kent, A. (1958). *Punched Cards, Their Applications to Science and Industry.* New York; Reinhold

Costello, J. C. (1962). 'Computer Requirements for Inverted Coordinate Indexes.' *Am. Docum.* **13**, No. 4, 414

Elsner, H. (1961). 'Erfahrungsaustausch "Handlochkarten in der Dokumentation".' *Nachr. Dokum.* **14**, No. 1, 53

Elsner, H. (1963). 'Erfahrungsaustausch mit Randlochkarten in der Dokumentation.' *Nachr. Dokum.* **14**, No. 1, 53

Engelhardt, H. (1958). 'Dokumentation in der Praxis.' *Schweiz. Med. Wschr.* **88**, H39, 960

Fugmann, R. (1962). 'Ordnung—oberstes Gesetz in der Dokumentation.' *Nachr. Dokum.* **13**, 3

Grolier, E. de (1962). *A Study of General Categories Applicable to Classification and Coding in Documentation.* Paris; UNESCO

Hauser, B. and Herrlich, K. H. (1969). 'Die maschinelle Herstellung von Sichtlochkarten grösserer Kapazität.' In: *Die ZMD in Frankfurt am Main.* pp. 172–178. Berlin; Beuth-Vertrieb

Heinze, H. (1957). 'Rentabilitätsbetrachtung einer Schrifttumskartei.' *Nachr. Dokum.* **8**, No. 3, 125

Hermann, P. and Löschner, G. (1961). 'Sichtlochkartei und gleichwertige Grundbegriffe zur Aufschliessung der Literatur über Diffusion.' *Dokumentation* **8**, No. 3, 125

Herrlich, K. H. (1963). 'Handlochkarten als Hilfsmittel der Zeichnungsarchivierung.' *Reprographie* **3**, No. 6, 127

Jans, W. (1956). 'Das Zeichenlochverfahren in der medizinischen Befunddokumentation.' *Nachr. Dokum.* **7**, No. 3, 134

Jolley, J. L. (1959). 'Punched feature card.' *Data Processing* **1**, No. 2, 86

Jolley, J. L. (1963). 'Mechanics of Coordinate Indexing.' *ASLIB Proc.* **16**, No. 6, 161

Jonker, F. (1959). 'The Descriptive Continuum: A "Generalized" Theory of Indexing.' *Proc. Int. Conf. Scient. Inf. Washington*

Kent, A. (1971). *Information Analysis and Retrieval.* New York; Becker and Hayes

Kistermann, F. (1957/58). 'Zur Geschichte und Entwicklung des Sichtlochkartenverfahrens.' *Dokum. Fachbibl. Werksb.* **6**, 7

Kistermann, F. and Uhlein, E. (1957). 'Die Sichtlochkarte.' *Umschau* **12**, 370

Loosjes, T. P. (1957). 'The Deltacard.' *ASLIB Proc.* **19**, No. 5, 142

Loosjes, T. P. (1964). 'Over doorzichtponskaarten.' *Bibliotheekleven* **49**

Manual (1955). *Document Reproduction and Selection.* The Hague; F.I.D. Publ. No. 264

Müller, K. H. (1970). 'Rationalisierung der Dokumentation durch und beim

Sichtlochkartverfahren.' In: *Probleme der Information in der Landwirtschaft.* Berlin; Deutsch Akademie der Landwirtschaftswissenschaften

Nelson, A. M. (1958). *Research and Development of the Magnacard System.* Office of Tech. Serv. U.S. Dept. of Commerce

O'Connor, J. (1962). 'The Scan Column Index.' *Am. Docum.* **13**, No. 2, 204

Peakes, G. L. (1957). 'Experiences with the Unit Card System for Report Indexing.' *Advances in Documentation and Library Science.* Vol. 2, pp. 306–327. New York; Interscience

Perry, J. W. and Kent, A. (1957). *Documentation and Information Retrieval.* Cleveland; Western Reserve University Press

Perry, J. W., Kent, A. and Berry, M. M. (1956). *Machine Literature Searching.* New York; Wiley

Plankeel, F. H. (1960). 'Automation in Documentation.' *Am. Docum.* **11**, 128

Proceedings (1959). *Proc. Int. Conf. Scient. Inf. Washington*

Report (1960). *Documentation, Indexing and Retrieval of Scientific Information.* Report of Committee of Government Operations. U.S. Senate 24/5/1960

Richens, R. H. (1958). 'Abstracting and Information Service for Plant Breeding and Genetics.' In: Casey, R. S. *et al.* (eds.). *Punched Cards.* pp. 374–387. New York; Reinhold

Scheele, M. (1959). *Die Lochkartenverfahren in Forschung und Dokumentation mit besonderer Berücksichtigung der Biologie.* Stuttgart; Schweizerbart'sche Verlagsbuchhandlung

Schüller, J. A. (1963). 'Het TDCK-Compact System.' *Tijdschr. Eff. Docum.* **33**, 11

Schüller, J. A. and Koekkoek, J. M. (1962). 'The TDCK-Compact System.' *J. Docum.* **18**, No. 4, 176

Shera, J. H., Kent, A. and Perry, J. W. (1956). *Documentation in Action.* New York; Reinhold

Shera, J. H., Kent, A. and Perry, J. W. (1957). *Information Systems in Documentation.* New York; Interscience.

Taube, M. (1953–1964). *Studies in Coordinate Indexing.* Washington; Documentation Inc.

Taylor, H. (1915). *U.S. Patent No. 1, 165, 465,* 1915 [Selective device]

Thompson, L. S. (1961). 'Feature Cards.' In: *State of the Library Art.* Vol. 4, Pt. 2, pp. 57–102

Vickery, B. C. (1961). *On Retrieval System Theory.* 1961. London; Butterworths

Vickery, B. C. (1970). *Techniques of Information Retrieval.* London; Butterworths

Weger, A. de (1959). 'Bedrijfsdocumentatie.' In: *Het Verantwoordelijk toegankelijk maken van literatuur.* pp. 9–14. The Hague; NIDER Publ. No. 24, 2nd series

Westendorp, J. (1956). 'De toepassing van ponskaartsystemen.' *Bibliotheekleven* **41**, No. 10, 289

Westendorp, J. (1959). 'Uber das Widerauffinden von nach dem Sichtlochkartenverfahren eingetragener Literatur.' *Nachr. Dokum.* **10**, No. 1, 24

Whaley, F. R. (1958). 'Retrieval Questions from the Use of Linde's Indexing and Retrieval System.' *Proc. Int. Conf. Scient. Inf. Washington.* Vol. 1, pp. 763–770

Whaley, F. R. (1961). 'The Manipulation of Non-conventional Indexing

Systems.' *Am. Docum.* **12**, 101
Wildhack, W. A. and Stern, J. (1957). 'The Peek-a-Boo System in the Field
of Instrumentation.' In: 'Information System in Documentation.' In:
Advances in Documentation and Library Sciences. Vol. 2, pp. 209–226.
New York; Interscience
Williams, W. F. (1966). *Principles of Automated Information Retrieval.*
Elmhurst, Ill.; The Business Press

EIGHTEEN

SPECIAL PROBLEMS IN SYSTEMS WITH POST-COORDINATION

Systems with post-coordination give rise to a number of problems, which will now be dealt with briefly. It must be emphasised that, within the terms of reference of the present work, this series of problems is not dealt with exhaustively and the treatment of individual problems is not meant to be conclusive but only indicative, leading the reader who wishes to pursue the matter further to the literature. The four problems to be discussed are:

1. Means of circumventing the limitations imposed by the surface area of the information carrier.
2. Extras.
3. Programming (key punching, coding, processing).
4. Automation.

MEANS OF CIRCUMVENTING LIMITATIONS IMPOSED BY SURFACE AREA OF INFORMATION CARRIER

(a) Freeing of Fixed Fields

Freeing of fixed fields can happen when the term entry is no longer fixed to a set field—that is, when the so-called column sequence is freed. This means that a certain coding can be entered *anywhere* on a card and still be retrievable by the mechanism. This would be an exceedingly economic exploitation of the card surface and would offer a possibility of introducing new codings, so long as space for this remained.

All this irrevocably means, however, mechanisation of the complicated 'retrospective search'. Thus a certain amount of space can be gained, at the cost of mechanisation and the considerable financial efforts involved, without fundamental changes.

Examples of punch cards without fixed field have been described by Garfield (1954). In this system the retrieval mechanism seeks out cards with a certain punch pattern, and it is immaterial where the card is punched.

With those devices employing transparent material such as microfilm an electronic eye senses out specific item entries corresponding to a given subject. The Rapid Selector worked with a filmstrip, half of which was the text (mostly title and abstract), the other half containing the subject index entry. In automated searching the speed was raised considerably; the machine, even at full speed, was able to work with a second filmstrip taking exposures of the passing frames which showed the target item entry. The Rapid Selector itself has become a museum piece; there are now various models on the market built on the same principle. Filmorex works with film shots which are divided as in the Rapid Selector but are in single frames and sorted by photo-electric scanning. By means of a drastic cut in the codification in the index entry field the constructor of the Filmorex, Samain (1947), has been able to limit the complexity of his machine, so that its price is within the budget of a normal documentation service. In addition, the discrete form of the units possesses some advantages over the filmstrip; against this, naturally, there must be a certain loss of retrieval speed. In addition, in Filmorex there is no fixed field, since Samain developed this system from an automatic punch card system without fixed field. Variants on this system are, for example, the Minicard and the microfilm selector of ERA (Engineering Research Association), which can search 20 codings simultaneously.

All the above are valid for Type IIB. In the case of IIIA the Uniterm card is in part also without fixed field and therefore possesses a comparatively high capacity. With Types I, IIA and IIIB fixed field is the only possibility.

(b) Superimposed Coding

With superimposed coding (SC), codifications being placed one on top of the other in the same part of the information carrier, the surface area of the carrier can be exploited more intensively, but the length of the descriptors must be statistically determined, so that the percentage of 'extras' may be kept within certain bounds. Calculations of this type have been made by Stiassny (1960) for various SC methods. Superimposed coding can be carried out in specific fields and also on the whole card, in which case, however, the fixed field is discarded. Superimposition does save space but takes a longer search-time because of the increased number of extras. A clear exposition of SC is given by Faden (1959).

This method has proved itself with edge-punched and slotted cards; it is even useful with machine-sortable punch cards. An example of this is l'Unité System of Te Nuyl (1958), who superimposes seven descriptors in one field as a totality.

Uhlmann (1964) gives a description of SC with superimposable punch cards and also discusses the possibility of combination with 'chain spelling'. In principle, in this connection, such a possibility is also a question of combination punching (see below).

(c) Combination Punching

This can be carried out by use of various keys (see p. 162).

(d) Other methods

Two further forms of these techniques for space saving are used with superimposable punch cards (see Chapter 17).

'EXTRAS' IN USING POST-COORDINATE SYSTEMS

In post-coordination systems queries can be posed to publications about A and B and T but not about $B:A:T$ or $B:T:A$. Therefore all too frequently many other documents are also retrieved, so-called 'extras'; they do not match up to the query (that is to say, not to the permutation sought but only to the combination as such). These 'extras' arise from the fact that the mutual relations of the descriptors are not fixed in the system and therefore a certain combination comes out irrespective of the permutation. Automatic systems only answer to a combination, the form of permutation in which the query is put to the combination not affecting the search.

The 'extras' which are obtained in this way can be placed in several categories according to cause:

1. The absence of a hierarchic reference. 'Influence of climate on afforestation': this phrase contains the descriptors 'climate' and 'afforestation'. Publications on this will, however, also be

found if the query is posed on the 'Influence of afforestation on the climate'. This can be avoided by introducing afforestation (subject) and afforestation (object). Such modifications are required also in the cases of homonyms, e.g. dog (mammal) and dog (instrument).

2. Compound terms disintegrate into elements and thereby cause incorrect combinations (false drops). Example: 'Use of sand in the preservation of potatoes' has the descriptors 'sand', 'preservation', 'potatoes'; but a publication on this subject will be retrieved also if the query on the preservation of sand-potatoes arose.

(3) The interchange of adjectives belonging to particular nouns. Example: a document on steel elements in timber construction would be found if timber elements in steel construction were sought.

Against the great advantage of not having to edit the permutation there remains the relatively minor disadvantage of 'extras' which have to be sorted out afterwards. A measure to prevent extras is the splitting up of the publication into small units which are dealt with as single documents (see Chapter 14). An attempt has also been made to employ indicators, which show the presence of a compound (and, if possible, the mode of this compound) between two terms. In the literature these indicators are described as role indicators, operators, links, etc. The process is also called interlocking or interfixing. A survey is to be found in Chapter 3 of the work by de Grolier (1962) and a standardisation of compound indicators was proposed by Farradane (1955, 1961). Artandi and Hines (1963) have described various modifications to the original Uniterm system involving more and more indicators.

PROGRAMMING (KEY PUNCHING, CODING, PROCESSING)*

The problem of programming arises only with unit cards (Type I) and item entry systems (Types II A/B).

There are almost as many programmes as there are users of item entry systems. For each case optimal programming is completely different. The main body of the literature on this problem consists of descriptions of these multifarious programmes, in which full rein is given to personal inventiveness. In this field no general advice

---

* *Translator's note:* Continental 'programming' can mean key punching, coding, processing, but not programming in the technical sense of writing a programme.

can be given; each individual case must be considered as a separate problem.

It is difficult to discover a pattern in these programmes. Naturally, the general aim is to obtain, with the minimum of trouble and material, the maximum amount of information. The effort is, however, constrained, since, the further it is pursued, the more difficult punching and read-back becomes.

It is generally considered uneconomic to use the so-called direct key, i.e. the simple and direct equation of one term, one punching. The flexibility of the system being used can be increased by combination punching (the so-called short key), the need for apertures being reduced accordingly. One of the most useful short keys is the 1–2–4–7 key, or, depending on the card system used, the 0–1–2–4–7 key, in which, by means of combination punching, the numbers 1 to 10 can be coded to allow sorting.

What more sophisticated keys allow is reported by Zschokke (1968, personal communication) for notched cards:

one needle	30
two needles	435 (with 1–2–4–7)
three needles	1000
four needles	11 025
six needles	91 125

It is particularly uneconomic to use *letters* in relation to language since (a) frequency of use of individual letters varies considerably in languages; (b) frequency differences are further exaggerated by the fact that certain combinations of letters (for example, English *th*) appear consecutively. A survey of the various keys and codes can be found in the textbooks mentioned at the beginning of Chapter 17.

Garfield (1961) applies principles of information theory to programming and shows that 25 per cent of the descriptors require 90 per cent of the work of punching and searching. Simple programming for the most-used descriptors, and increasingly complicated programming for descriptors the less they are used, is Garfield's advice. Vickery (1960) has indicated that a certain standardisation in the field of descriptor and machine languages should be attempted, so that interconvertibility could be achieved between the 'languages' at present in use.

AUTOMATION

The choice of post-coordinate systems is often made when retrieval criteria should actually have made pre-coordination the system of choice. The reason is the fact that if systems are being automated for retrieval, they are usually post-coordinate systems. There are then two main grounds for automation:

1. High storage capacity with small volume and machine processing. If, for example, the documents were to be split up into thought units, there would scarely be more than three descriptors per unit; but the number of documented units rises correspondingly faster and, as a result, automation of storage and retrieval is chosen.
2. Possibility of automation of ancillary tasks, such as sorting-in, retrieval, print-out and eventually counting. In countries with high labour costs it is most advantageous to have such tasks as sorting-in, retrieval, print-out and counting automated. In addition, it is often felt that repetitive tasks (which are boring in themselves) are too time-consuming and the inclination is to install a machine, although in reality the 'lost time' by no means justifies the use of a machine.

Another psychological motive for introducing machines can be the (unconscious) desire to appear progressive. Shaw (1951) and Perry (1950) specifically mention the fact that machines are mostly constructed for other purposes—especially punch card machines, devised in the first place for administrative purposes.

The decision to automate a system for reasons outlined above alters nothing in principle, although the machines can carry out certain retrieval steps much more quickly, especially in cases of searching with queries displaying the structure of a logic sum or difference (see p. 167).

It is unnecessary here to pursue the matter of automation and the various machines available, since there are excellent textbooks in this field—for example, Kent (1962), Bourne (1963) and Becker and Hayes (1963). Furthermore one can keep up-to-date with the reports of the numerous conferences.

Holm (1962) surveys the automated documentation systems currently available; Gull (1962) gives details of where calculating machines have been applied to documentation processes in the U.S.A.; and Salton (1964) surveys activities in Western Europe.

Critical assessment of what measure of help the documentalist may expect from machines can be found in Mills (1964) and de Grolier (1961). Vickery (1961) agrees as to automation of clerical operations but doubts the greater efficiency of machines in the information retrieval phase, which must be designated 'non-clerical'.

He says (p. 151):

Let me now try to sum up the situation as I see it. There is a good case for mechanizing the acknowledged clerical operations which occur in retrieval systems, particularly in constructing files from index entries, and in delivering search results. There is sometimes a good case for mechanical file searching, but the possibly greater efficiency of human search must not be overlooked. The potential benefits of automatic indexing (formulating entries) are great, so that the problem is worth exploring, despite the great difficulties it presents. On the other hand, there is a danger of over-mechanization. It is always necessary to consider whether a simpler machine can be used by reorganizing the process, for example, by changing the file organization. Lastly, we must make sure that advantages claimed for a machine system are in fact inherent in the machine, and not just brought about by an improvement in the system, e.g. in subject coding, which could be applied manually with equal effectiveness.

References

Artandi, S. and Hines, Th. C. (1963). 'Roles and Links—or Forward to Culture.' *Am. Docum.* **14**, No. 1, 74

Becker, J. and Hayes, R. M. (1963). *Information Storage and Retrieval.* New York; Wiley

Bourne, C. P. (1963). *Methods of Information Handling.* New York; Wiley

Faden, B. R. (1959). 'Information Retrieval: Punch Card Techniques and Special Equipment.' *Spec. Libr.* **50**, No. 6, 244

Farradane, J. E. L. (1955). 'The Psychology of Classification.' *J. Docum.* **11**, No. 4, 187

Farradane, J. E. L. (1961). 'The Challenge of Information Retrieval.' *J. Docum.* **17**, No. 4, 233

Garfield, E. (1954). 'Preliminary Report on the Mechanical Analyses of Information by Use of the 101 Statistical Punch Card Machine.' *Am. Docum.* **5**, No. 1, 7

Garfield, E. (1961). 'Information Theory and Other Quantitative Factors in Code Design for Document Card Systems.' *J. Chem. Docum.* **1**, No. 1, 70

Grolier, E. de (1961). 'Considérations sur d'une part le choix des systèmes mécaniques de recherche d'information en fonction de leurs conditions d'emploi; d'autre part les caractéristiques désirables des systèmes de codification.' *ADIA Proceedings, Beiheft Nachr. Dokum.* No. 8

Grolier, E. de (1962). *A Study of General Categories Applicable to Classification and Coding in Documentation.* Paris; UNESCO

Gull, C. D. (1962). 'Automatic Documentation, Current Systems and Trends in the U.S.A.' *Revue Docum.* **29**, No. 2, 57

Holm, B. E. (1962). 'Searching Strategies and Equipment.' *Am. Docum.* **13**, No. 1, 31

Mills, J. (1964). 'Information Retrieval. A Revolt Against Conventional Systems?' *ASLIB Proc.* **16**, No. 2, 48

Perry, J. W. (1950). 'Information Analysis for Machine Searching.' *Am. Docum.* **1**, No. 3, 133

Salton, G. (1964). 'Automatic Information Processing in Western Europe.' *Science, N.Y.* **144**, 627

Samain, J. (1947). 'Progrès du classement et de la sélection mécanique des documents.' *F.I.D. 17th Conf. Bern. 1947. Rapports* 1, pp. 22–6

Shaw, R. R. (1951). 'Management, Machines and the Bibliographic Problems of the 20th Century.' In: *Bibliographical Organization.* pp. 200–235. Chicago; Chicago Univ. Press

Stiassny, S. (1960). 'Mathematical Analyses of Various Superimposed Coding Methods.' *Am. Docum.* **11**, 155

Te Nuyl, Th. W. (1958). 'L'Unité Documentation System.' *Revue Docum.* **25**, No. 3, 65

Uhlmann, W. (1964). 'The Application of Random Superimposed Coding and Chain Spelling to Peek-a-Boo Cards.' *Am. Docum.* **15**, 89

Vickery, B. C. (1960). 'Coding for Interconvertibility.' In: Kent, A. (ed.). *Information Retrieval and Machine Translation.* Vol. 2, Ch. 54. New York; Interscience

Vickery, B. C. (1961). *On Retrieval System Theory.* London; Butterworths (2nd edn 1965)

NINETEEN

THE CHOICE OF RETRIEVAL SYSTEM

In Chapters 11 to 18 various factors influencing the choice of the retrieval system have been discussed. To recapitulate:

1. Influence of choice of document analysis in three respects: (a) size of unit chosen for documentation (Chapter 15); (b) degree of compression (after condensation) chosen (Chapter 11); (c) number of descriptors per unit documented and the related question of free permutation (Chapter 15).
2. Influence of choice of the thesaurus in two respects: (a) degree of built-in structure (Chapter 12); (b) endeavour to keep the thesaurus as short as possible (p. 106).
3. Choice in emphasis on work investment: at either input or output stage (Chapter 15).
4. Choice in relation to predictability of queries as well as to the subject field, to hierarchic structure, or to the situation of a central problem or central object of study with properties or influencing factors (Chapter 12).
5. Freedom of development of the systems in the fields of the number of descriptors and of the number of documents, respectively (Chapter 16).
6. Short cuts taken in the stages of document analysis→retrieval system (Chapter 14).
7. Choice of systems with post-coordination because of automation, on grounds other than retrieval (p. 163).

The following additional factors, which also influence the choice of the retrieval system, are now dealt with:

1. Structure of the queries and search strategy.
2. Evaluation of retrieval systems.
3. Further cost–benefit management factors

166

STRUCTURE OF THE QUERIES AND SEARCH STRATEGY

In Chapter 9 one aspect of the search habits of research workers has been examined : namely, the kind of sources they use for their information and the order in which these sources are used. The search habits of research workers can also be studied in connection with specific retrieval systems. Six types of problem arise :

1. Hierarchic structure of the queries.
2. Logical structure of the queries.
3. Browsability optima.
4. The possibility of retrieving documents which only partially or marginally contribute something to the object of search.
5. Difference in 'factual value' of descriptors.
6. Building-in of successes as 'feed-back' in the systems.

Hierarchic Structure of the Queries

As previously discussed, the choice between document analysis without thesaurus or with thesaurus is dependent upon the pre-dictability of the queries. The choice of degree of hierarchy is dependent upon the hierarchy of the queries. The degree of hierarchy determines the freedom in permutation: the less hierarchy, the more permutation and vice versa. As has been shown, the choice between systems with post- or pre-coordination is governed by the number of descriptors and their permutations. The number of descriptors is influenced by the factors listed in Chapter 15. The number of permutations and the circumstances, too, under which this number changes are both dealt with in Chapter 15.

Finally, reference should be made to Vickery (1965, Chapter 7), where various aspects of this problem are dealt with in greater depth.

Logical Structure of the Queries

In practice most queries possess the structure of a logical product, something about A in relation to B and also to C, or, in UDC language, $A:B:C$. It sometimes happens that $A:B$ or C is requested; that is, $A:B$ or $A:C$ (logical sum). It is also possible that $A:B$ or $A:C$ might be asked for, but not $A:D$ (logical difference). The frequency of occurrence of queries of the last-named type depends apparently upon the situation. Herner and Herner (1958) have found, as a result of their statistical work in this field, that 98 per cent of the queries had the structure of a logical product, 1.2 per cent that of a logical sum and 0.8 per cent that of a logical difference. The corresponding figures of Whaley (1958), however, are 21.4 per cent, 63.8 per cent and 1.4 per cent. The remaining 13.4 per cent consisted of queries with only one descriptor. Fugmann (personal communication) has said that, particularly in the treatment of queries of logical sum or difference type, the machine is much faster than a manual search. However, in queries of logical product type, the advantages of the machine are open to doubt.

For detailed discussion on the above-mentioned structures, reference can be made to, for example, Kistermann (1962) and Shera, Kent and Perry (1956).

Browsability Optima

There is always the danger that librarians and documentalists may erect barriers between the user and the literature, which perforce direct the queries of the user to well-worn customary paths. Now the documentalists cannot be asked to extract from the literature something which only later might possibly turn out to be a brand-new subject heading; otherwise they might as well take over the research themselves. The topics for which scholars are searching may often lie between two subject headings or between two sets of facts. Wise and Perry (1950) have formulated it thus: 'No amount of human skill in devising indexes and classification can, however, anticipate future trends in viewpoint and in research. Yet it is precisely in the direction of the unexpected correlation and the surprising result that the most spectacular progress is made.'

What is it that actually disturbs the reader? He misses in the retrieval system that browsability which he himself develops if he is allowed to rummage around in a collection.

One might compare a searcher in the retrieval system to a motorist in a strange town having to 'presort' the route without fully knowing where the choice lies, as he is not familiar with the lay-out of this strange town. The searcher in a library may known only vaguely what he wants and must find it by heuristic processes. That means, however, that one must visit the library to view the literature which remains *in situ*, non-loanable and arranged according to subjects. All this is a question of space and money since:

(a) A document on two subjects must be duplicated.
(b) The individual documents are often so disparate in form that they scarcely fit together in a bay—for example, thick books, thin offprints, microcards, microfilm and so on.
(c) Freely accessible library rooms are more expensive than store-rooms or archives.
(d) The documents are not separately produced but bound together—for example, in issues of periodicals.

It is to be hoped that reproduction processes will, at some future time, make it possible, within the bounds of a normal budget, to duplicate indefinitely: for double or several shelf positions, for loan, for separation of volumes supplied bound together, and so on. At present this is not the case. In the future it is conceivable that, instead of the card catalogue, the microcard (film or opaque) with the integral text could take its place. In addition, the punch card on diazo material, Filmorex or Microcite systems and electrostatic processes could all play a part. Kent (1971) has described still other possibilities, including the use of magnetic ink.

Ideal conditions for the researcher today can only be found in small institute libraries, where the offprint collection is usually in class order and can be freely scanned; where non-loanability or restricted loanability largely prevails; where the material is more or less the same in outer form; and where, perhaps, duplicates have been prepared for duplicate storage. Nevertheless, for the future, such an arrangement of bibliographic material and retrieval systems can be hoped for that the searchers have a certain degree of freedom to browse, if not among the documents themselves, at least in the bibliographic material. It is necessary first of all to be clear about the work methods of research and the way in which searchers can best be provided with retrieval systems in the bibliographic material.

In the interests of documentation, it is important to establish the work pattern of the research worker. Researchers do not bring specific queries (excepting the cases when an orientation in a marginal area is sought) as do visitors to a public library. Workers using literature have, it is true, adapted themselves more or less to librarians' and documentalists' systems and catalogues; they are looking, however, for 'suggestions, parallelisms and stimulations,' as Berry and King (1959) have expressed it. Perry (personal communication) said that the best work method for a research chemist in the course of an investigation is to 'alternate' between library and laboratory, so that the work advances well-coordinated on both fronts.

Menzel (1964) provides a good survey of the diversity of queries to be expected from researchers; it is clear that in view of this variety a unified universal system cannot be the answer.

In general, it can be said that the greater the desired freedom the retrieval systems allow the searcher, the more it costs in search-time; the scanning of many entries in a KWIC index or in the index to *Chemical Abstracts* is time-consuming but gives the browser considerable freedom. On the other hand, systems with, for example, a rigid classification reduce the search work and similarly the amount of freedom.

The Possibility of Documents which either Partially or Marginally Contribute Something to the Object of Search

Lancaster (1964) enumerates six groups of subsidiary documents which can be called upon in the event of insufficient documents being found which correspond to the descriptors of search:

1. Documents cited by other documents. This has already been dealt with in the 'snowball' system in Chapter 9. To cite does not mean that the whole document is of interest, but only a paragraph or even a sentence.
2. Documents citing other retrieved documents. This can be done by means of 'Citation Indexes' (see p. 84), with the same reservation as in (1) above.
3. Documents with index term profiles 'similar' to profiles of other retrieved documents. 'Similar' means, therefore, that not all but some of the descriptors should be the same (see also under 'Browsability optima').
4. Documents having one or more citations in common with other retrieved documents. This is what Kessler (1963) calls 'bibliographic coupling'. The strength of the 'coupling' depends upon the number of citations in common.
5. Further documents by authors of retrieved documents. This is a normal search technique and as such requires no further clarification.
6. Documents retrieved by terms statistically associated with original search terms. These terms refer to statistics of association, already dealt with in Chapter 12 on the construction of descriptor lists. It is the use of the machine which makes it possible, within an acceptable time limit, to pursue many 'association trails' (Kent, 1971), where a searcher in card catalogues would have given up much earlier. It is necessary

that these 'trails' be also built into the programme, just as *see* and *see also* are in the usual subject indexes.

Difference in 'Factual Value' of Descriptors

The factual value becomes important in searching with n descriptors when the results are incomplete and search with $n-1$, $n-2$ descriptors, etc., must therefore be pursued. Which descriptor should be dropped first (generic level) and so on? The building-in of a 'weighting' factor into the system is recommended by, e.g., Bar Hillel (1960). Schultz and Shepherd (1961) stress in addition the great difference in retrieval value of descriptors and they suggest hints for an optimal choice of descriptors by means of numbers.

Building-in of Successes as 'Feed-back' into the System

Feed-back is possible with machines and imitates the storage of positive results in the brain of the literature searcher, but with greater freedom from error. In this case the feed-back acts exactly as does the human memory.

EVALUATION OF RETRIEVAL SYSTEMS

We have already discussed in Chapter 12 the evaluation of the performance of descriptor languages—more specifically the recall/precision method. We now refer to some important work on system evaluation.

One of the most prominent authors in this field is Lancaster (1968a, b; 1969), whose diagram of the system parameter network we reproduce (*Figure 34*). Lancaster first worked with Cleverdon in the Cranfield experiments (mentioned in Chapter 12) and he later applied similar evaluation methods in the U.S.A. to a test of MEDLARS (Lancaster, 1969) which turned out with 58 per cent recall and 50 per cent precision. Seven items for improvement were indicated.

In the U.S.A. the Comparative Systems Laboratory of the Case Western Reserve University was active for several years. The 3-volume report of this laboratory was the object of an elaborate discussion by Vickery (1969) and Richmond (1970).

A valuable review on evaluation of systems is that of Cleverdon

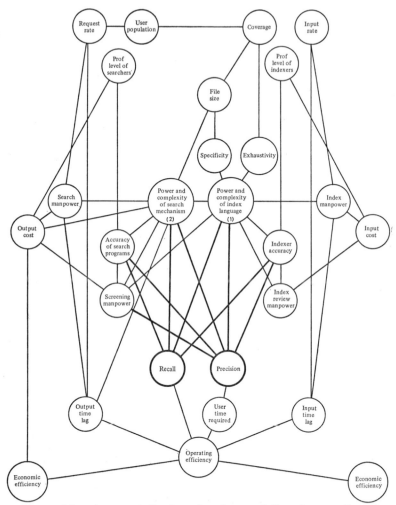

Figure 34. Diagram indicating the factors influencing recall and precision: 1, Specificity, automatic generic posting, weighting, links, roles. 2, Making available at the same time number, title and abstract. (From Lancaster, F. W. (1968a), by courtesy of Journal of Documentation. *Notes 1 and 2, and some thickening of lines, added by author)*

(1970). He discusses at length the experiments as far as they have been carried out in an environment of complete information storage and retrieval systems. Seventeen different series of experiments are reviewed and the main conclusions are: (a) experiments have been successful; (b) techniques are increasingly efficient; (c) work has concentrated on the field of indexing and indexing languages to the neglect of the output and of the system covering the actions and reactions of users; (d) recall and precision are still the best criteria of effectiveness. The areas in which the author sees possible future developments are: (1) quality control of retrospective searches; (2) comparison between various agencies as inaugurated by Leggate; (3) justification of IR services on a cost–benefit basis.

If a more intensive study of Cleverdon's methods within broader terms of reference is projected, the work of Swets (1963) is recommended. He compares 10 different methods of efficiency measurement (Cleverdon's among them) and adds a suggestion for an eleventh system.

The question under review is actually a question of time— namely how much time one is prepared to spend in re-sorting the 'extras' in order to make the extraction of information as exhaustive as possible.

FURTHER COST–BENEFIT MANAGEMENT FACTORS

A general treatment of the problem of efficiency evaluation of documentation systems can be found in, for example, Herner (1962).

'Apparatus' for mechanical retrieval of item entry cards is so expensive that, in most cases, it would not be a paying proposition for scanning and evaluating the literature. Automatic systems, then, in most cases are considered only if literature retrieval can run simultaneously as a subsidiary process, but it is just this fact which gives rise to delays. This is the substance of a complaint from industrial practice by Bayer at Leverkusen. Gray (1950) established that, for research workers, depth of analysis, up-to-dateness and percentage of cover are more important than the costs thereby incurred.

Kollberg (1956) has compared the costs of a classical vertical card system and those of an edge-punched card system. The heaviest cost factor in the classical card system is the sorting and filing. In contrast to the classical card system, the punch cards save 50 per cent in space and 27 per cent in costs of material.

Thorne (1955) has in the same way carried out costings for Uniterm, UDC and edge-punched cards.

Hirayama (1962) presents detailed cost comparisons between normal legible card, edge-punched card and centre-punched card. A survey of costs of various systems with post-coordination is given by van Dijk (1963). Stevens (1961) has published a cost comparison between systems with pre-coordination and systems with post-coordination.

As already mentioned (Chapter 15), in post-coordination the input costs are lower but the output costs are higher. The reverse situation is found in systems with pre-coordination. Overmeyer (1962, 1963) gives detailed data on input and output costs of a fully automated and operational search service for metallurgists.

Numerous publications dealing with running costs can be found in *Nachrichten für Dokumentation*, mainly as a result of the efforts of a committee on cost–benefits of the German Society for Documentation.

Gull (1956) has published an extensive comparative examination of expenditure in time and money of different methods in the input stage of a Uniterm system. Since the differences in cost were not particularly large but differences in speed were (the slowest requiring three times as long as the fastest), it is the speed factor (together with other factors specific to a case) which apparently will have the heaviest weighting. Even so, it is the author's opinion that efficiency is appreciably less important in this stage than it is in the analysis stage, since it is this stage that is always the weakest point. The maximum speed of the retrieval mechanisms should not be confused with the sorting speed, whose top rate is often much lower. A sorting machine can often be blocked by a single query; while in the classical card system, with all its other boxes or drawers, other queries can be simultaneously satisfied.

Danilof (1957) has compared systems on the basis of various management costing criteria (search-time and others). This very interesting investigation showed that the preparatory time in classical card systems, edge-punched cards and slotted cards is approximately equal; in centre-punched cards it is about 25 per cent faster. Comparisons of search-time for one term result in the following relations between, respectively, classical card systems, edge-punched cards, slotted cards and centre-punched cards: 26–14–10–5. For two terms: 35–12–9–1. For three terms: 70–10–7.5–5.

Zschokke (1968, verbal communication) compared the cost of notched cards and slotted punch cards, finding the latter group three to four times as expensive but giving five times as much opportunity for input of combinations as notched cards. Shaw (1951) expressly draws attention to the errors in the estimation of running costs which can result from the use of different machines for work in literature retrieval. For example, preparation is counted but not the tasks following the mechanical work. Often the only correct solution is a combination of manual and automatic work. As an example of this, Shaw (1951) mentions an experiment in the alphabetical arrangement of 10 000 cards. This work took:

manually only	106 h
machine only	56 h
combinations of both methods	23 h

In practice, manual and automatic work is often combined when the presort (préclassement) is carried out manually, as at Rhone-Poulenc (Paris) and Centre National de la Recherche Scientifique (Samain), where the punch cards are stored presorted in categories. In addition, the extras are said to be manually culled after automatic sorting. Jonker (1958) esteems the speed of the system so highly that he classifies systems with post-coordination according to the speed of processing.

A further possible element in evaluation of running costs is, for example, the question whether a system could be automatically duplicated, which is of importance in the decentralisation of an organisation. The cost in time and money for programming alone can be a decisive factor. The same applies in automation, which may well lead to greater speed but also entails certain constraints. For example, it may be necessary to know whether the searchers can search the system themselves (as in a card catalogue) or whether they must finally delegate search to a programmer or technician (browsability). It is, of course, quite possible to learn how to browse with the aid of a machine (see p. 168). Also of importance is the number of queries the machine can deal with simultaneously, as compared with a card catalogue, for example, in using which a searcher blocks only *one* drawer. The possibilities of weeding might also prove to be of great importance.

The so-called 2nd Cranfield Conference (Proceedings 1970) was entirely devoted to the question of the cost of information storage and retrieval systems.

SOME CONCLUSIONS

Although the conclusions below are partly based on scarcely tenable generalisations, it can still be said that, in many cases, preliminary guidance can be given. A person charged with the responsibility for choosing a retrieval system cannot avoid a deeper analysis of his own work, as discussed earlier in the consideration of possible criteria. *There is no panacea.* Only those cases where the choice between systems with pre- and post-coordination has already been decided in favour of the latter are mentioned here.

Workers within a discipline and research workers or small research teams specialising in one field should in the first instance consider the following systems edge-punched: slotted card, term entry system with numeric comparison and term entry system with comparison by superimposition. The term entry system with numeric comparison is so very time-consuming that it is only applicable to pilot systems (for choice of descriptors), before one changes over to the system of comparison by superimposition. The main difference between edge-punched and slotted cards on the one hand and term entry systems on the other, from the viewpoint of running costs, is the fact that much more time must be devoted to 'input' in the case of edge-punched and slotted cards. Against this, term entry systems need more time in the 'output' stage, since only numbers are to be found there—not, in the first instance, the sought-for references (which in the other systems can be written in at the same time onto the edge-punched or slotted cards). The references, therefore, must be given a number (addressed) and a list of these numbers must be kept (or the documents themselves arranged by *numerus currens*). Edge-punched and slotted cards are mostly employed with fixed fields.

For medium-large documentation centres or services with an annual intake of at least 20 000 items, manual methods do not suffice; a modicum of automation is required. For this reason consideration can be given to centre-punched cards with fixed fields, for which simple sorting machines suffice; or unit cards, which may well impose higher machine demands but are much more flexible in construction and, above all, can also be used without programming.

For very large documentation services, automated item and term entry systems without fixed fields may be suitable; to these can be added, perhaps, the unit card system, although the large number of cards used can quickly lead to difficulties. It is more probable that documentation will here be split up into smaller units, grouped according to some useful characteristic.

References

Bar Hillel, Y. (1960). 'Some Theoretical Aspects of the Mechanization of Literature Searching.' *Technical Rep.* No. 3. U.S. Office Naval Research Contract No. N62558–2214

Berry, M. M. and King, G. W. (1959). 'Summary of Area and Discussion.' *Proc. Int. Conf. Scient. Inf. Washington*

Cleverdon, C. (1970). 'Evaluation Tests in Information Retrieval Systems.' *J. Docum.* **26**, No. 1, 55

Danilof, H. (1957). 'Vergleichende Zeitstudie verschiedener Verfahren zur Dokumentation von Patentschriften.' *Nachr. Dokum.* **8**, No. 4, 179

Dijk, M. van (1963). 'Essai sur le coût de la recherche documentaire par les méthodes d'indexation coordonnée.' *Revue Docum.* **30**, 143

Gray, D. E. (1950). *Study of Physics Abstracting. Final Report.* New York; American Institute of Physics

Gull, C. D. (1956). 'Posting for the Uniterm System of Coordinate Indexing.' *Am. Docum.* **7**, No. 1, 9

Herner, S. (1962). 'Methods of Organizing Information for Storage and Searching.' *Am. Docum.* **13**, No. 1, 3

Herner, S. and Herner, M. (1958). 'Determining Requirements for Atomic Energy Information from Reference Questions.' *Proc. Int. Conf. Scient. Inf. Washington.* Vol. 1, pp. 181–187

Hirayama, K. (1962). 'Time Required, Cost and Personnel for Documentation.' *Am. Docum.* **13**, No. 3, 319

Jonker, F. (1958). 'Design Considerations of Information Storage and Retrieval Machines.' *ASTIA Doc. AD154273, Air Force Off. Scient. Res. Centre* No. AF 49(638)91

Kent, A. (1971). *Information Analysis and Retrieval.* New York; Becker and Hayes

Kessler, M. M. (1963). 'Bibliographic Coupling Between Scientific Papers.' *Am. Docum.* **14**, No. 1, 10

Kistermann, F. (1962). 'Fragenstrukturen.' *Arb. Bl. Betr. Informationswesen* (ABI) **44**, No. 66, 67

Kollberg, R. (1956). 'Steilkartei und dreireihige Randlochkarte im Vergleich.' *Nachr. Dokum.* **7**, No. 2, 81

Lancaster, F. W. (1964). 'Mechanized Document Control.' *ASLIB Proc.* **16**, 132

Lancaster, F. W. (1968a). 'Evaluating the Economic Efficiency of a Document Retrieval System.' *J. Docum.* **24**, No. 1, 16

Lancaster, F. W. (1968b). *Information Retrieval Systems.* New York; Wiley

Lancaster, F. W. (1969). 'Medlars: Report on the Evaluation of its Operating Efficiency. *Am. Docum.* **20**, No. 2, 119

Menzel, H. (1964). 'The Information Needs of Current Scientific Research.' *Libr. Q.* **34**, No. 1, 4

Overmeyer, L. (1962). 'Test Program for Evaluating Procedures for the Exploitation of Literature of Interest to Metallurgists.' *Am. Docum.* **13**, No. 2, 210

Overmeyer, L. (1963). 'An Analysis of Output Costs and Procedures for an Operational Searching Service.' *Am. Docum.* **14**, No. 2, 123

Proceedings (1970). '2nd International Conference on Mechanized Information Storage and Retrieval Systems.' *Inf. Storage Retrieval* **6**, Nos. 1 and 2

Richmond, Ph. A. (1970). 'The Final Report of the Comparative Systems. Laboratory: A Review.' *J. Am. Soc. Inf. Sci.* **21**, 160

Schultz, C. K. and Shepherd, C. A. (1961). 'A Computer Analysis of the Merck, Sharp and Dohme Research Laboratories Indexing System.' *Am. Docum.* **12**, No. 2, 83

Shaw, R. R. (1951). 'Management, Machines and the Bibliographical Problems of the 20th Century.' In: *Bibliographical Organization.* pp. 200–225. Chicago; Univ. of Chicago Press

Shera, J. H., Kent, A. and Perry, J. W. (1956). *Documentation in Action.* New York; Reinhold

Stevens, N. D. (1961). *A Comparative Study of Three Systems of Information Retrieval.* New Brunswick; Grad. School of Libr. Science. Rutgers State Univ.

Swets, J. A. (1963). 'Information Retrieval Systems.' *Science, N.Y.* **141**, 245

Thorne, R. G. (1955). 'The Efficiency of Subject Catalogues and the Art of Information Searches.' *J. Docum.* **11**, No. 3, 148

Vickery, B. C. (1963). 'Vocabularies for Coordinate Systems.' *ASLIB Proc.* **15**, No. 6, 170

Vickery, B. C. (1965). *On Retrieval System Theory.* London; Butterworths

Vickery, B. C. (1970). *Techniques of Information Retrieval.* London; Butterworths

Whaley, F. (1958). 'Retrieval Questions from the Use of Linde's Indexing and Retrieval System.' *Proc. Int. Conf. Scient. Inf. Washington.* Vol. 1, pp. 763–770

Wise, C. S. and Perry, J. W. (1950). 'Multiple Coding and the Rapid Selector.' *Am. Docum.* **1**, No. 2, 76

INDEX

Italic numbers indicate location of bibliographical details.

Textual form, 104
'Thesaurus', 95, 97, 110, 112, 121, 166, 167
Thompson, L. S., *156*
Thorne, R. G., 174, *178*
Tornüdd, E., 70, *87*
'Trails', 171
Translation, of the condensate, 127
Triangular layout, 149, 152
Tydeman, J. F., 9, *20*

UDC, 5, 111, 115, 127, 128, 133, 134, 135, 174
Ühlein, E., 153, *155*
Uhlmann, W., 160, *165*
Ulrich's International Periodicals Directory, 47
UNESCO, 23, 41, 48, 54, 57
Union catalogue, 57, 59
Union List of Serials, 50
UNISIST, 41, 121
Unit card, 145, 146, 161, 176
Unité system, 160
Uniterm, 106, 122, 132, 151, 159, 161, 174
Urquhart, D. J., 31, *45*, 77, 82, *87*
User's approach, 93, 96

Varossieau, W. W., 35, *45*, 53, *55*
Velander, E., 83, *87*
Vensenyi, P. E., 99, *100*
Verdoorn, F., 48, *52*
Verein Deutscher Bibliothekare, 18
Verhoef, M., 1, *8*
Verkerk, P., 49, *52*
Verzeichnis Schweizerischer Zeitschriften, 47
Vickery, B. C., 39, *43*, 99, *100*, 105, *109*, 110, 116, 117, 119, 121, 122, *125*, 127, *128*, 135, 136, *141*, 147, *156*, 162, 163, *165*, 167, 171, *178*
Visscher, M. B., 78, *87*
Vital Notes, 50

Wadsworth, R. W., 54, *55*
Watson, J. D., *74*, *87*

Watson Davis, 5
Webb, E. C., 16, *20*
Weeding, 175
Weger, A. de, 135, *141*, 153, *156*
Weil, B. H., 107, *109*
Weinberg Report 6, *8*, 21, *26*
Weiss, P., 56, 57, *60*
Welch Medical Library Projects, 49
Wells, A. J., 122, *124*
Welt, I. D., 107, *108*
Werzig, G., 105, *109*
Wesseling, J. C. G., 122, *125*
Westendorp, J., 152, 153, *156*
Whaley, F. R., 147, 148, *156*, 168, *178*
Whitford, R. H., 48, *52*
Wildhack, W. A., 152, 153, *156*
Williams W. F., 94, *100*, 147, *156*
Willing's Press Guide, 50
Winchell, C. M., 25, *26*
Wise, C. S., 168, *178*
Woke, P. A., 19, *20*, 57, *60*
Wolk, L. J. van der, 64, 65, *68*
Wood, D. N., 70, *87*
Wood, G. C., 37, 38, *45*
Wood, J. L., 28, *45*
Woodford, A. O., 12, *19*, 35, *43*
Word association coefficient, 113
Word 'clumps', 113
Word frequency, 106
World Economic Agricultural and Rural Sociology Abstracts, 32
World List, 22, 49
World Medical Periodicals, 11, 48, 49
Wright, W. E., 30, *45*
Written formulae, 95
Written profile, 91
Wyatt, H. V., 36, *45*, 96

Yale Medical Library, 81
Yngve, W. H., 129, *131*

Zarember, I., 107, *109*
Ziman, J. M., 23, *26*
Zipf–Estoup curves, 74, 84, 85
Zschokke, H., 162, 174